Dexippus

On Aristotle's *Categories*

Dexippus

On Aristotle's *Categories*

Translated by
John Dillon

Cornell University Press
Ithaca, New York

First published 1990 Cornell University Press.

Library of Congress Cataloging-in-Publication Data

Dexippus, the Platonist.
[In Aristotelis categorias. English]
On Aristotle's categories / translated by John Dillon.
p. cm.
Translation of: In Aristotelis categorias.
Includes bibliographical references.
ISBN 0-8014-2266-3
1. Aristotle. Categoriae. 2. Categories (Philosophy) — Early works to 1800. I. Dillon. John M. II. Title.
B438.D4913 1990
160—dc20 90-33372

Printed in Great Britain

Contents

Introduction

Dexippus' life and works

Under this heading there is unfortunately little to be said.[1] From Dexippus himself we learn only that at the time of composing the present work he had a daughter about whose health he was anxious (or even whose death he may be mourning), and that he presided over a philosophical school, of which the otherwise unknown Seleucus was at least one of the star pupils. That Dexippus himself was a student, or at least a philosophical partisan of Iamblichus, can be deduced from Simplicius' description of him in the introduction to his *Commentary on the Categories* (p. 2,9 Kalbfleisch) as 'the Iamblichean' (*ho Iamblicheios*), and from the fact that Iamblichus dedicates a letter to him on Dialectic, a passage from which is preserved by Stobaeus (*Anth.* II 18,12-19,11 Wachs.), but he is not mentioned among the pupils of Iamblichus by Eunapius (*Lives of the Sophists and Philosophers*, p. 458 Boissonade) though that is an open-ended list (Eunapius speaks of 'many others').[2] Certainly he is no amateur philosopher. His work shows a thorough acquaintance with technical terminology, and a good understanding of the philosophical issues, at least by the standards of his time.

The present work is the only one he is known to have

[1] The entry in *RE*, composed by Kroll, vol. 5, cols 293-4, summarises what there is, as does Zeller in *Phil. d. Gr.* 3: 2^3, 736-7.

[2] It is curious, however, that Eunapius, in celebrating the *historian* Dexippus, just prior to dealing with Iamblichus (p. 757), describes him as 'a man overflowing with culture and logical power' (*paideias te kai dunameôs logikôs anapleôs*), but this Dexippus, since he led a force against the Herulians who occupied Athens in AD 267 and defeated them, is a generation too early to be a follower of Iamblichus. However, he could, I suppose, have had a son of the same name or even a grandson, who inherited his *dunamis logikê* – unless Eunapius is simply confusing the two Dexippi, as he is quite capable of having done, and assumes that the *Categories* commentary was written by the historian; cf. A. Busse, *Hermes* 23, 1888, 402ff.

written, and it would seem reasonable to date it some time in the first decades of the fourth century AD (Iamblichus himself died in the early 320s). Dexippus is making use, on his own admission (5,9), not only of the commentary of Iamblichus on the *Categories*, but also of those of Porphyry – the big *Commentary to Gedalius* (now lost), and perhaps also the surviving short commentary in question-and-answer form, which now precedes his own commentary in the *CAG* Series.

The nature of the commentary

Like Porphyry's short commentary, Dexippus' work is in dialogue, or rather, question-and-answer form, though he makes considerably more effort to give his work literary form than does Porphyry.[3] The origins of this form of commentary are obscure. It has an ancestor in the 'problems and solutions' type of commentary, of which Philo of Alexandria's *Questions and Answers on Genesis* and *Exodus* are an example, and more remotely in the philosophical dialogue generally, but there are no surviving examples of what one might term the 'catechetical' commentary before that of Porphyry. It doubtless dramatises a school practice of which we know something from such sources as Aulus Gellius' *Attic Nights* and Porphyry's *Life of Plotinus*, the master's making himself available for questions.[4] The connexion with the 'problems and solutions' type of commentary is made by Dexippus himself – indeed at 5,10ff., when he is explaining how he is not going to compete with his immediate predecessors, Porphyry and Iamblichus, both of whom have composed vast and comprehensive commentaries, he makes it clear that he will confine his treatment to 'the disputed questions' (*ta aporoumena*).

He does indeed do that, making use (no doubt at second hand through Porphyry and Iamblichus), first of the earlier,

[3] Apart from the introductory passages to the three books, which indulge in some literary flourishes and quotations of Hesiod and Pindar, Seleucus is allowed to develop something of a personality throughout the work, and there are a few lively exchanges. By contrast, Porphyry's interlocutors are quite anonymous and impersonal.

[4] Gellius *NA* I 26; Porph. *VP* 13. Cf. also Eunapius' account of Iamblichus fielding questions in *VS* 460.

Middle Platonic and Stoic critics of the *Categories*, such as Eudorus, Lucius, Nicostratus, Atticus, Cornutus and Athenodorus, and then increasingly, from Book 2 on, of objections raised by Plotinus in *Enneads* 6.1 and 3, and possibly (again *via* Porphyry) in oral communication.[5]

The content and extent of the commentary

Dexippus' declared purpose, as we have seen, is to cover all the 'problems' that have been raised over the years against the *Categories*, and from a comparison with Porphyry's surviving commentary, and with that of Simplicius, we can see that he does do that fairly comprehensively.

The criticism of the *Categories* had a long history, going back to the Platonist Eudorus of Alexandria (*fl. c.* 25 BC) in the first generation after Andronicus' edition, and continuing, as we have just noted, in both the Platonist and Stoic schools down to Plotinus, who criticises both Aristotle and the Stoics in *Enn.* 6.1.[6] The tradition was derivative and cumulative, each later author simply taking over his predecessors' *aporiai* (questions), amplifying them (or alternatively, summarising them), and adding a few of his own. On the evidence of Simplicius (deriving ultimately from Porphyry), the most comprehensive collection of *aporiai* seems to have been that of 'Lucius and Nicostratus', that is to say, as he makes clear at *in Cat.* 1,19, Nicostratus taking over Lucius (*ta tou Loukiou hupoballomenos*).[7] Nicostratus was a Platonist philosopher of the mid-second century AD,[8] so that Lucius may be dated, presumably, to the first or early second century. Simplicius'

[5] See on this possibility Paul Henry, 'Trois apories orales de Plotin sur les *Categories* d'Aristote', in *Zetesis* (Festschrift E. De Stryker), Antwerp/Utrecht 1973, 234-65. While conceding the possibility that a number of *aporiai* mentioned by Dexippus are based on Porphyry's memories or notes of Plotinus' seminars rather than on the text of the *Enneads*, I am unable to work up much excitement about the addition to our knowledge of Plotinus' thought that this involves.

[6] The whole tradition is now excellently covered in volume 2 of Paul Moraux's great work, *Der Aristotelismus bei den Griechen,* Berlin 1984, 509-601.

[7] We have fully 26 references to this pair (or one or the other of them) in Simplicius' commentary.

[8] He is honoured by the Delphians in an inscription to be dated somewhat before AD 163 (Ditt. *Syll.* II^3, no. 868), along with three other (otherwise unknown) Platonists. See Karl Praechter, 'Nikostratos der Platoniker', *Hermes* 57, 1922, 481-517, and Paul Moraux, op. cit., 528-63.

description of Nicostratus' procedure is important for our understanding of the tradition which Dexippus is inheriting, as it explains the frequently fiddling and even absurd objections which we see him raising – or allowing Seleucus to raise (*in Cat.* 1,18ff.):

> Others have chosen to confine themselves solely to raising problems against the text, as is the case with Lucius and after him Nicostratus, who took over what Lucius wrote, making it their business to bring objections against pretty well everything written in the book, and not with any discrimination, but recklessly and shamelessly. We may be grateful to them, however, both because many of their objections do raise important (*pragmateiodeis*) issues, and because they have given those who came after them opportunities, arising out of the solution to the problems, for many fine theoretical developments.

We can see, indeed, from the surviving testimonies, that the practice of Lucius and Nicostratus was to fire at anything that moved, but, as Simplicius says, interesting issues are often raised in the process.

In the late second century, Atticus continued the tradition, but seems to have added little to it, and Plotinus, in his turn, bases himself very largely on Lucius and Nicostratus, though applying his characteristic penetration to the problems.

On the Stoic side, the earliest known critic is a certain Athenodorus who wrote a work *Against Aristotle's Categories* (Porph. *in Cat.* 86,27). Moraux[9] argues for his identification with Athenodorus, son of Sandon, of Cana (near Tarsus) in Cilicia, who lived in the last half of the first century BC. Whether or not the identification is correct, this is certainly the period when Athenodorus must have written his work, since he is utilised, and criticised, by his fellow Stoic L. Annaeus Cornutus in the mid-first century AD.[10] Cornutus, a teacher and friend of the poet Persius, is the author of an extant *Critique of Hellenic Theology* and some works on rhetoric, but he concerned himself also with criticism of the *Categories*, composing a *Reply to Athenodorus* in which he

[9] *Aristotelismus* 2, 585-7.
[10] See Moraux, *Aristotelismus* 2, 592-601.

criticised some of his positions while plainly adopting others. Of these, at least the Platonist tradition was known to Plotinus (though perhaps only through the writings of Atticus), but the learned Porphyry was acquainted with all of them, and it is through him, in his big Commentary dedicated to Gedalius, in seven books, that their work is passed on to the later Neoplatonic tradition beginning with Iamblichus.[11]

What we find in Dexippus, then, is a boiled-down version of the results of Porphyry's erudition, probably largely mediated through Iamblichus (though there are a few instances, adverted to in the notes,[12] where he seems dependent on Porphyry in opposition to Iamblichus). In most cases, it is only by using the vast commentary of Simplicius as a control that we can see what is going on. Simplicius knows of Dexippus' existence (*in Cat.* 2,25ff.):

> And Dexippus, the follower of Iamblichus (*ho Iamblikheios*) also wrote a concise commentary on the book of Aristotle, but primarily sets out to give solutions to the problems raised by Plotinus,[13] which he presents in dialogue form, adding almost nothing himself to the work of Porphyry and Iamblichus.

However, it seems most improbable that he used him as a source. In many cases, indeed, when they are verbally very close, Simplicius specifies that he is actually following Iamblichus.[14] Dexippus may be regarded, then, as an independent witness to Iamblichus' commentary, and a series of passages, where he is verbally very close to Simplicius, may be claimed with fair certainty for Iamblichus' commentary (which would not, in most of these cases, it must be said, have differed much from that of Porphyry).

[11] Simplicius is explicit (*in Cat.* 1,9ff.) that Iamblichus followed Porphyry very closely, even verbatim, in most of his commentary, merely introducing qualifications and clarifications at certain points, but also bringing Archytas into the argument, and overall giving a distinctive 'intellectual interpretation' (*noera theôria*) of the subject matter. Dexippus does take note of Archytas on occasion (16,33ff.; 65,8ff.), but he has not much use for the *noera theôria*.

[12] 26,9ff. (n. 85); 43,10ff. (n. 20); 65,14ff. (n. 7).

[13] This is true, really, only for the portion of the commentary from Book 2 on. Book 1 concerns *aporiai* from the earlier tradition.

[14] = Dex. 17,25-9ff.; 61,19-62,6 = Dex. 39,10-40,5; 99,4-9 = Dex. 49,2-9; 100,13-101,12 = Dex. 49,26-50,24; 106,28-107,4 = Dex. 51,23-52,4; 130,8-19 = Dex. 69,6-25; 131,10-16 = Dex. 70,9-14.

Dexippus, then, is of interest both as a surviving testimony to the great achievement of Porphyry in turning aside Plotinus' rejection of Aristotle's *Categories* and as partial evidence for the content of Iamblichus' commentary,[15] but is there anything distinctive that we can claim for him? Probably, if we possessed the commentaries of Porphyry and Iamblichus *in toto*, we would find very little that was original to Dexippus, but as things stand, there are a few cases where we may accord him the benefit of the doubt.

The first instance is his extended treatment of the arguments presented by the Peripatetic philosopher Sosigenes[16] on the subject matter of the *Categories* (7,4-9,22). There is nothing corresponding to this in Simplicius' nor in Porphyry's short commentary, so that it looks almost as if Dexippus has done this bit of research on his own. From the extended treatment which he gives it, one might conclude that he was rather proud of himself. He seems to have got hold of a sort of dialectical exercise composed by Sosigenes to sharpen the wits of his pupils (to adopt Moraux's suggestion, op. cit., 338) and is using it as a basis for developing his own view, that the true subject matter is *noêta*.

This is perhaps his most notable contribution, but there are other, smaller ones, probably less original. One is his presentation of some problems raised by Plotinus against the *Categories* which do not appear (either at all, or so fully) in *Enneads* 6.1 or 3. Fr. P. Henry has devoted a long article to these,[17] dealing chiefly with chs 2.8 (an *aporia* on substance), 3.7 and 3.11 (two *aporiai* on *logos*), and showing, with great

[15] In this connexion, it must be said that B.D. Larsen's collection of the fragments of that Commentary in his *Iamblique de Chalcis: Exégète et Philosophe*, Aarhus 1972, vol. 2, though useful, is seriously incomplete, since he confines himself to assembling the passages explicitly attributed to Iamblichus by Simplicius, taking no account of the parallel passages in Dexippus, which reveal many places where Simplicius is using Iamblichus without attribution, simply because he has no disagreement with him. A good example, perhaps, of the value of Dexippus as a surviving link in the exegetical tradition is the passage 40,19-42,3, where he first (40,19-25) gives what is probably Porphyry's argument against Plotinus that the *Categories* is not a bad attempt to talk about *things*, because, being designed for beginners, it is instead about *words* as significative of things; and then (41,18-42,3) presents the (Iamblichean?) explanation of how Aristotle's account of substances can be applied, if one chooses, to the intelligible Forms, because Forms are neither said of, nor present in, a subject.

[16] On whom see Paul Moraux, *Aristotelismus* 2, 335-60.

[17] cf. above, n. 5.

probability, that Dexippus is relaying material presented originally by Porphyry from records of oral criticisms by Plotinus of the *Categories* which amplify what we have preserved in *Enneads* 6.1 and 3. As I have said above (n. 5), while interesting, this supplementary material does not come across as particularly sensational. In 3.8, Plotinus[18] makes the objection that Aristotle does not give a positive definition of Substance in *Cat.* 3a8-9, but merely a description by negatives ('neither said of a subject nor in a subject'). Plotinus actually adumbrates this objection in *Enn.* 6.1, 2.15-18 ('But in general it is impossible to say what Substance is: for even if one gives it its "proper characteristic" (*to idion*), it does not yet have its "what it is" (*ti esti*)'), but he certainly does not develop it as it is developed here. I see no difficulty in supposing that Porphyry is amplifying Plotinus' objection here from his own personal memories or notes of what his master used to say. Similarly in the case of 3.7: Plotinus at the beginning of 6.1.5 raises, in his usual compressed manner, the problem about language (*logos*) that it is really only as sound, or more exactly, sounded *air*, that it is a *quantum*. Language as such, though, is rather an impact (*plêgê*) on air, and so, if anything, is an instance of 'making'. What we find in Dexippus and Simplicius is an amplification of this, no doubt, but essentially the same point, and Henry seems to me to make rather too much of it.

3.11 is the most interesting of the three passages with which Henry deals, as it seems to correspond to nothing in *Enn.* 6.1-3. Unfortunately, Dexippus' text has given up by this time, so we have only the summary:

> That language, being a thing involving combination (*sumplokê*), cannot fall under the categories, at least if the categories are simple items. (cf. *Cat.* 1a16; 1b25)

Simplicius mentions the *aporia* briefly (*in Cat.* 130,32-3), attributing it, not to Plotinus, but, vaguely, to 'they' (*aporousin*), which I would take as more naturally referring

[18] If we accept that it *is* Plotinus. The *aporia* is not explicitly identified as his either by Dexippus, or by Simplicius (91,15-32), though it is sandwiched between two Plotinian ones, and the manner of its introduction in Dexippus certainly suggests that it is by the same author as its predecessor. Also, *to idion* of Dex. 44,9 probably echoes the same term in *Enn.* 6.1.2,17.

either to Lucius and Nicostratus, or to the whole critical tradition. Henry regards it as 'certain' (op. cit., 243) that this is an *aporia* of Plotinus, since it comes at the end of a series of *aporiai* about *logos*, all the previous ones of which are identified as emanating from him. I am afraid that I cannot share his certainty. Both Dexippus' summary and Simplicius' had the opportunity so to identify it if they chose, and they do not. I do not see that it can be claimed confidently as an oral *aporia* of Plotinus, as opposed to an *aporia* which was just floating around in the tradition, and picked up by Porphyry.[19]

Other than these oral *aporiai*, there are just a few features of interest. The last *aporia* of Book 1 (1 40), which occurs neither in Simplicius nor in Porphyry and thus may be Dexippus' own contribution, seems to present a distinctive notion of existence, arising out of the sophistic problem that to say substance, etc. *exists* is either redundant, or creates a parallel set of *existent* categories, which could then be said in turn to exist, producing an infinite regress. Dexippus counters this by denying that adding 'is' indicates anything distinct from the subject, but merely bears witness to its subsistence (*hupostasis*).

There are a number of other passages in the commentary which find no parallel in Simplicius or Porphyry: 1 7; 1 14; 1 23; 1 32; 1 35; 1 39; 2 1; 2 4;[20] 2 9;[21] 2 17-19; 2 26; 2 32; 3 24; 3 29-30; 3 37. None of these problems are, I think, of any great importance. They are generally just scholastic elaborations of other problems which do figure in the tradition, and some of them are very trivial indeed. We cannot be sure that Simplicius did not find them in Porphyry and/or Iamblichus and decide to omit them, but there could be instances of Dexippus venturing to add a little to the tradition himself.

It is plain, then, that no great claims can be made for

[19] We may also note in this connexion 2 14, where Dexippus attributes to Plotinus an *aporia* which is reported also in Simplicius (89,30-90,6), no echo of which appears in the *Enneads*. Henry does not actually deal with this passage, though it would have helped his case. It seems to me possible, however, that Porphyry (to whom all this goes back) simply tacked this *aporia* on to the previous one, which is itself not Plotinian, but attributed by Simplicius only to 'some people' (*tines*).

[20] A Plotinian *aporia* not mentioned by Simplicius, so it may be the result of Dexippus' own reading of *Enneads* 6,1-3, though one would assume Porphyry (and Iamblichus) to have dealt with it as well.

[21] Once again, no trace of this *aporia* in Simplicius, so perhaps an original contribution.

Dexippus as an original thinker, nor is there any suggestion that he aspired to be one, but in view of the fragmentary state of the tradition of commentary on the *Categories*, it can be seen that his little work takes on a certain importance.

As regards the original extent of the commentary, there seems no reason to suppose that Dexippus did not intend to cover the whole *Categories*, including the *Postpraedicamenta*, which he regards as part of the work (*in Cat.* 17,7-9). Simplicius seems still to have had access to the whole work, so the present mutilation of it took place later than his time. As we have it, the work covers only the text up to 4b23, in the middle of the discussion of Quantity, which the preserved table of contents of the third book shows to have been concluded in that book. The whole work may have been somewhat more than three times its present size, if we may judge from the fact that the present text corresponds to 132 of the 438 pages of Simplicius' commentary in the *CAG* edition.

The present translation

Doing justice to the technicalities of Dexippus' philosophical vocabulary and, at times, compressed argumentation, has not been easy. I have tried to keep the needs of modern philosophers in mind by maintaining consistency of technical vocabulary as far as possible, and by liberal insertion of the original Greek terms in brackets. This is clumsy, as are a number of the terms I have been forced to use, but, to appreciate the scholastic points that are being made, it is essential to know exactly what terms are being used. A common item (*koinon*) or 'commonality' (*koinotês*), for example, is not a universal (*katholou*), nor is a definition (*horos*) an account (*logos*) – though I have sometimes rendered the latter as the former, when nothing substantive is at stake. A *hupokeimenon* is sometimes a subject, and sometimes a substrate. Subsisting (*huphistasthai*) is distinct from existing (*einai*, *huparkhein*); and so on.[22] I have been greatly aided in

[22] Strictly speaking, *koinon* or *koinotês* should refer to a common property, as a broader concept than a universal (e.g. 45,24; 53,10-15); a *logos*, again, is a looser concept than *horos*; 'subsisting' is traditionally applicable to a broader range of entities than 'existing' (at least in the sense of *huparkhein*); and *hupokeimenon* has to serve both as a physical and as a grammatical 'subject'. But I must confess that the

this, and other matters, by the assiduous comments of Professor Steven Strange, whose translation of Porphyry's short commentary will appear in this series. The full extent of his help is only imperfectly acknowledged in the notes, as is that of another, anonymous reader.

I have decided to include the chapter headings in the text of the translation, rather than grouping them at the beginning (where they are to be found in the manuscripts), or omitting them altogether, in order to help the reader by providing both divisions and indications of the subject matter. The chapter headings are probably not by Dexippus himself, which is why they are placed in square brackets, but they are reasonably authoritative,[23] and sometimes help to elucidate the text. I have also included the *personae* of the dialogue in each case, though the manuscripts omit them.

The text of the commentary

The standard text of Dexippus is that of Adolf Busse in the *Commentaria in Aristotelem Graeca* (*CAG*) Series of the Berlin Academy (1887), in which it figures as vol. 4, part 2. The first edition of the work is in fact no older than 1859, that of Leonard Spengel, published in Munich by the Bavarian Academy.[24] Much earlier than an edition of the Greek text, however, there appeared a Latin translation, by Felix Felicianus of Verona, published in Venice in 1546 by Hieronymus Scotus. Though based on an inferior manuscript (*Coislianus* 332), this translation occasionally contributes useful insights.[25]

Busse lists ten manuscripts of the work, all of which go back to a single late medieval exemplar, which was itself mutilated, and four of which he considers worthy of attention. These he divides into two families, one comprising the manuscripts C (*Laurentianus* 72, 21, of the fifteenth century),

real differences between these terms as used by Dexippus in any given instance is frequently less than clear to me.

[23] On a few occasions they are less than accurate, e.g. 1 25, 2 26, but generally they hit the spot.

[24] This edition was reviewed by Hermann Usener, in the *Litter. Centralblatt* 1860, 124-5, and he offers some useful emendations.

[25] on 12,7; 15,7-8; 33,14; 54,7-8.

and R (*Parisinus gr.* 1942, of the fourteenth century), the other comprising A (*Laurentianus* 71, 33, fourteenth century) and M (*Matritensis* 76, fourteenth century), the CR tradition being the better of the two. I have been content to accept his text, save in the instances noted in the list of emendations or variant readings. The evidence of Simplicius in parallel passages is frequently useful, but sometimes problematical, as we shall see.

Dexippus

On Aristotle's *Categories*

Translation

Dexippus

Commentary in dialogue form on Aristotle's *Categories*[1]

BOOK 1

Prologue

SELEUCUS: A certain diffidence takes hold of me as I make bold to ask a most fair and noble favour of you, Dexippus, my instructor in all that is good, but since I am mindful of Hesiod's dictum (*WD* 317): 'Modesty is no good companion of a man in need', I am encouraged in my desire to discuss this matter. 4,5

DEXIPPUS: But I for my part have long wondered at you, most excellent Seleucus, that, though still so young, you are so conspicuous in aptitude for learning and gentleness, you attack your studies vigorously, and retain what you have learned firmly and unerringly, while under the stimulus of mental agility and sharpness you jump ahead of everyone else in your questions, and now I am ready to accede to your enthusiastic request for instruction. For you are lacking in none of the gifts of good fortune, and in addition you are in point of breeding and reputation equalled by none of your contemporaries. I commend your natural readiness for any noble purpose, and the fact that you despise all external goods, but love education and are insatiable for learning. So, as regards whatever philosophical topic you wish, pray command me! 10 15

SELEUCUS: Turn your attention, then, to the problems that 20

[1] The full title in the mss. is 'Dexippus the Platonic Philosopher, Problems and Solutions in relation to Aristotle's *Categories*, Book the First. The work is in dialogue form, and the characters in the dialogue are Dexippus himself and Seleucus.'

have been raised in relation to Aristotle's *Categories*, and try to solve the disputes that have arisen.

DEXIPPUS: A tall order, indeed, my dear and valued friend, and not at all easy for me to fulfil at the present juncture. For
25 both the misfortune[2] which has befallen my clever and beautiful daughter Diiphile seems grievous and troublesome
5,1 to me (why should one not speak the truth?) and my body is wasted with diseases. I am glad, then, on the one hand, to accede to your request, but I have resolved to agree from afar, as do cowardly contestants, because it is hard to contradict the Platonic philosopher Plotinus when he has produced such penetrating difficulties. However, for your sake I will not
5 shrink from going forth, like the Homeric hero, 'even against a divinity'.[3] But do you aid me in the argument by not insisting on detailed exegeses; for many scholars, in particular Porphyry, and then later Iamblichus, have produced a vast number of commentaries on this work, difficult to master
10 because of their bulk. So then, that I may not fall into the same situation as them, please confine your questions to the disputed questions. For it is not my ambition to fill any deficiencies in their treatments (I do not flatter myself so much), but I simply wish to provide solutions which are swift and concise and clear. (Whoever is intending to read these problems should first of all make sure that he has clearly in mind the details of the text, since otherwise he will understand nothing of what is said here).[4]

[1. Why have many people raised problems about Aristotle's *Categories*?]

15 SELEUCUS: Well then, what is the reason that impelled the philosophers of old to engage in disputes of every kind with each other on the subject of this Aristotelian writing which we

[2] See Introduction, p. 7.

[3] Menelaus says this, at *Iliad* 17.104, when contemplating the onset of Hector. The implication is that, if Seleucus will play Ajax (whose aid Menelaus prays for in the *Iliad*), then the project might succeed. Another implication, perhaps not to be pressed too hard, is that Plotinus is *daimonios* or even *theios*. It is not clear how early this terminology of 'canonisation' for Platonic philosophers came in.

[4] This editorial note rather breaks the illusion of the dialogue. Seleucus ignores it. I take it to be a gloss.

call the *Categories*.[5] For as far as I can see, neither have more numerous controversies occurred about any other topic, nor 20 have greater contests been stirred up, not only by Stoics and Platonists trying to undermine these Aristotelian Categories, but even among the Peripatetics with each other, with some assuming that they have more perfectly grasped the man's meaning, while others think that they can solve with relative ease the problems raised by opponents.

DEXIPPUS: Because, most industrious Seleucus, the subject 25 of this book concerns the primary and simple utterances <and the things>[6] they signify. So since rational discourse (*logos*) is useful to all branches of philosophy, and the first principles of this are simple utterances and their objects of reference, it is natural that much controversy has arisen as to whether Aristotle has dealt correctly or incorrectly with the subject.

[**2**. Why is the book entitled 'Categories'?]

SELEUCUS: But what is the significance of his title, and the 30 name 'Categories'?[7] For he presumably doesn't intend to explain how men accuse each other in law courts; for the 6,1 Greek language has given this term to prosecution (*katêgoria*), to which the opposite is defence (*apologia*), so that one might reasonably accuse him of an unnatural use of terms.

DEXIPPUS: Many responses to this accusation have been made, Seleucus, on behalf of Aristotle, and these may be read 5 in the commentaries of Alexander and Porphyry. I propose to pass over most of them, and will instead set out that which seems to me to be the most penetrating, i.e. that names are not on an equal footing with things, nor do signs have the

[5] Criticism of the *Categories* goes back at least to Eudorus of Alexandria (fl. *c.* 25 BC) in the first generation after Andronicus published his edition. The history of controversy is well documented in Simplicius' massive commentary, and see Introduction, p. 9.

[6] I would supply *kai tôn pragmatôn* after *lexeôn*. The text as it stands makes no sense, and this supplement is supported by what follows just below.

[7] This whole discussion of the meaning of the title down to 6,23 is reflected in Simpl. 16,31-17,3 and 17,26-18,3 (17,4-25 is a report of Porphyry's view, followed by a comment by Simplicius), though it is not verbally close. Simplicius is probably closer to the Iamblichean original. The substance of the discussion is also to be found in Porphyry's short commentary (55,3-56,13).

same nature and status as the things signified, but the signifying power of names very greatly falls short (of the
10 complexity) of things. There is, then, every necessity for philosophers either to employ strange terminology quite distinct from ordinary speech, since they are exponents of matters unknown to the general public, or they must use ordinary speech, and make an extended[8] use of words originally coined for another purpose. For since names are established to be symbols and signs of things, it will necessarily be the case that
15 for things that are generally familiar there should already be names established which mean those things, while such things as are the objects of (specialised) knowledge demand the coinage of terms by specialists. Thus Aristotle too sometimes resorts to neologism in his imposition of terminology, as in the case of his creation of the word 'entelechy', while in other instances he transfers words from common usage to signify and
20 present a concept which he is engaged in expounding, as in the present case of the term 'category'; for whereas *katêgorein* is normally used to mean 'accuse under an indictment (*aitia*)', the philosopher has extended the meaning to a simple signification in speech that something is (predicated) of something else. So the title 'Predications' (*katêgoriai*) indicates that genera and
25 species and all universal terms are 'predicated' of those things which fall under them, and universally significant utterances are predicated of all the individual utterances which are ranked under them.

[3. What does a predicate signify – expression, thing or object of thought?]

SELEUCUS: Well, you have explained with excellent terseness that the '*Categories*' are so called from the verb 'to be predicated', that is, to be said of some subject; and indeed
30 we must take thought for conciseness above all else, lest we fall into the same boundless prolixity as our predecessors. But since I find what it is that is said (*to legomenon*) puzzling, whether it is an expression or a thing or an object of thought,[9]

[8] The word used is *metapherein*, but the meaning is rather broader than 'make a *metaphorical* use of'.

[9] The Greek terms are *phônê*, *pragma*, *noêma*. 'Expression' for *phônê* must be

pray endeavour to expound this to me clearly.

DEXIPPUS: I must first give a summary reply to your question: the primary object of reference is concepts, but incidentally also things. But since there has been a great deal of discussion about subject matter,[10] I too wish to speak about it at some length. Sosigenes the Peripatetic,[11] for example, set out a series of parallel enquiries about the subject matter, and yet did not come down decisively in favour of any one solution, but left the arguments locked in equal combat. I therefore want both to examine his arguments, and finally to bring forward my own view. 7,1 5

In putting forward the case of things, and wishing these to be the subject of statements,[12] he employs the following set of arguments:

(1) 'If things are somehow the decisive factor in relation to speaking or not speaking, so that if they do not exist, we say nothing, and if they exist we make a statement, they would be what is said.' 10

In reply to this I would say nothing prevents *things* from being the causes (*aitia*) of speaking, although they are not immediately (*prosekhôs*) *what is said*, which is what we now are seeking. For example, exposure to the sun is the cause of fever, and a prick of a needle the cause of pain, but neither is the needle *present* in the pain, nor the exposure to the sun *in* the fever. For, of causes, some are external and separate, 15

understood in the sense of 'spoken word'. *Pragma* will generally translate 'thing', but 'state of affairs', or 'fact' or 'reality' (a term I borrow from the French of Philippe Hoffmann, 1987) may be necessary to cover the whole range of meaning, and I shall use them also, when that seems appropriate (with transliteration of the Greek). The translator of *pragma* should take account of such key Aristotelian passages as *DA* 3.8, 431b25 (where Hamlyn translates *pragmata* as 'objects' (of perception or thought) and *Int.* 16a7 (where Ackrill has 'actual things'). Simplicius' discussion of the *skopos* of the work, in 9,4-13,26 of his Commentary, should also be consulted. As for *noêma*, it means 'concept' in a sense, but not in quite the same way as *ennoia*. It is, more strictly, an 'object of thought'.

[10] The subject matter, or *skopos*, of the treatise is discussed by Simplicius from 9,4-13,26, and the historic controversy between the advocates of words, things and concepts is set out.

[11] Sosigenes was a teacher of Alexander of Aphrodisias. See *RE* 'Sosigenes' (1), III A, 1157, and now P. Moraux, *Der Aristotelismus bei den Griechen*, II, 335-60. It is interesting that Simplicius makes no mention of Sosigenes or his parallel arguments.

[12] I choose this rendering here of *ta legomena*, but sometimes it seems best to render *legein* as 'utter', and *ta legomena* as 'what is uttered', since Dexippus uses the expression sometimes to mean 'the subject-matter of this treatise (sc. the *Categories*)', and sometimes simply to mean 'what is being said', or the subject of utterance.

while others are present in the effect, so that it is possible both for things to be causes of our utterance, but yet not to be what is uttered. For we speak about things that are not present before us, and about past and future things, and, since the statement about them exists, the thing would have 20 to exist also. But if the things are not called into existence with the utterance, they would not be what is said. For it is possible to speak about non-existent entities too, such as centaurs and goat-stags, and madmen and ecstatics speak; so that if in fact what is said were things, they too would be bringing their statements to bear on things, and we would not be able to name anything that did not exist.

25 (2) 'Again,' he says, 'if the truth or falsehood of a statement derives from things, it could not be the case that what is uttered is one thing, and what makes the truth or falsehood another; so that those things would be the subjects of statements which are causes of the complete statement.' In making use of this argument, I am surprised that Sosigenes did not realise that, if it were the case that every statement is true or false, he would have seemed to make sense in 30 asserting that it is things that are what is said, but if there is 8,1 a class of statement that is neither true nor false, and it is possible to make a statement which bears on no subject matter, how is it possible for him to maintain this argument? Further, it is one thing to speak about a thing, and another to make a true or false statement about it, but our present enquiry concerns the subject-matter of utterance. In addition, 5 even granting that the thing is the cause of a true or false statement, it does not necessarily follow that things are what is referred to; for it is possible that, if someone were to grant this, a statement might come to be true or false because of these (things), but the objects of reference of the statements might be different. For it is not the case that when a statement is receptive of truth or falsehood, it derives its 10 truth or falsehood from things; for being true or false is an accidental attribute of a statement, whereas its making a primary reference is essential to it.

(3) 'But,' he says, 'if,[13] even after being mentioned, things

[13] Reading *ei* after *all'*, to preserve the syntax.

(*pragmata*) such as 'horse', for example, or 'ox' or 'stone' continue to exist, it is obvious that it is these that are what is signified; for otherwise they would no longer exist when the utterance was completed.' But I would say just the opposite, that just in view of this things are not what is uttered, nor 15 does utterance have relation primarily to them. For the utterance which talks about them is transitory, while the things themselves remain. But if it were things that were what is signified, it would necessarily follow that they too should no longer exist when the utterance has passed away. As it is, however, one may have the case where someone is speaking and, as it might be, also listening (to another), and the things about which he is speaking do not exist. And yet if it is necessary that the agent should exist at the time when 20 the patient is receiving the action, it is necessary that what is uttered should exist at that time when the hearer is hearing. But if the speaker and the hearer exist, but the things do not, when the statement concerns past events, how is it possible for things to be what is signified?

Such, then, is our treatment of the case for things. But if one says (4) that it is the utterance that is what is being talked about (for after all many people can hear this and it can 25 exist and be spoken both when the things exist and when they do not), we shall say that it is one thing for something to be signified by something; for if we understand by speaking expressing something with the voice, then utterances are spoken, since they are expressed through the voice; but if naming is not just saying a name, but (saying) something else 30 by means of the name, then neither would saying be just uttering a sentence (*logos*), but rather (saying) something else by means of the sentence, so that now we are seeking what is the thing that is being signified through the statement, not 9,1 what we actually utter in speech.

(5) 'But,' he says, 'even as walking is an activity of a walker, and writing of a writer, even so speaking is an activity of the speaker, and the speaker goes through a sequence of words and parts of speech, so that it would be these that would be the objects of speech.'

But I would say that the activity itself is one thing, and the 5 result of the activity another; for the building is not the same thing as the activity of the builder, nor the painting the same as the activity of the painter, but in each case the results are distinct; so that the speaking, as being the result of an activity[14] of the speaker, has something else (outside itself) which is what is signified by the result of the activity.

(6) Again, he says that 'speaking' (*legein*) is formed paronymously from *lexis* or *logos*, so that on this ground also it
10 would result that it would be utterances (*lexeis*) that would be the objects of speech.

But this suggestion directly refutes itself. For if 'speaking' is formed paronymously from 'speech' (*logos*), we would not be speaking speeches, if it is true that we do not walk a walk or are healthy a health, nor that we do or create deeds or creations.[15]
15 Again, those who utter ambiguous or homonymous terms make one single utterance, but signify more than one thing, so that utterances would thus not be the things signified. Furthermore, when one makes an unclear statement, we hear the actual words, but we say that we do not understand what the speaker is saying, so that it would not be the words that are being said or signified, but something else. And what are we to say about meaningless words, when the utterance is
20 uttered, but signifies nothing? And again, in the case of polyonymous words, there are a multiplicity of words, but only one object (*pragma*), so that, if the object is not the same as the utterances, it would have to be something else than the actual things uttered (*ta legomena*).

This, then, completes this dialectical exercise. The ancients, however, declare that the only things signified are objects of thought (*noêmata*).[16] Since these are about things (*pragmata*)

[14] I take this to be the meaning of *energêma*, as opposed to *energeia*.

[15] This is true only if we bear in mind that what is being translated here is *praxeis* and *poiêseis*, words denoting processes rather than the results of processes. The general point is that the objects of our actions are not the actions themselves but the results of the actions.

[16] cf. Simpl. 41,28ff.: 'Boethus says that according to the ancients the only things uttered and signified are thoughts.' Since what follows in Simplicius follows Dexippus virtually verbatim for six lines, down to the beginning of the Aristotle quotation (which Simpl. refers to but does not quote), and since Iamblichus is quoted by Simplicius just above (41,20), I suggest that the common source here is Iamblichus' commentary, though Iamblichus will doubtless be borrowing from Porphyry.

I have altered Busse's punctuation here, connecting the 'since' clause with what follows, rather than what precedes.

and arise from things, it is objects of thought which are signified primarily, but on a secondary level things also; for 25 truth and falsehood do not reside in things, but in thought and the processes of the mind. At any rate, Aristotle, in Book 3 of his treatise *On the Soul*, speaks as follows (3.6, 430a26ff.): 'The thinking of individual objects is already a combination of concepts.'[17] So it is concepts, then, that are the primary 10,1 objects of signification and utterance; for concepts occur to us about non-existent objects, which we then interpret in speech. And true and false speech would not come about, if speech did not receive the truth and falsehood from concepts; for the 5 primary kind of speech occurs in the reasoning faculty, from which it is reasonable to assume that speaking itself and the uttered speech takes its name.

Again, ambiguities and homonyms and all deceptions which arise in speech come about by virtue of there being a multiplicity of objects of thought, since one's thought lights upon one or another of them; so if there is only one word involved, and the objects of thought connected with it are multiple, then also the things being said are multiple. 10

It is possible to prove the same point from the opposite case also. For if I say both *methu* and *oinos* for wine, there is only one concept corresponding to both of them, and yet the words are different and not the same. And further, meaningless words demonstrate with absolute directness that it is concepts that are the objects of reference; for we say that such and such a person is 'saying nothing', since it is impossible for either the speaker or the hearer to generate any concept on the basis of such utterances. Those who babble, then, are said 15 to be 'saying nothing' by virtue of the fact that they are conceiving of nothing. And in general no one would say that the other animals employ speech, plainly because they do not share in thought (*dianoia*). So it is plain to us from all these considerations that it is concepts that are signified primarily, and secondarily things.

A predication, then, is an utterance which supplies, for example, the concept (*noêma*) of 'Socrates', not differing, 20

[17] For Aristotle, the 'undivided objects' are the ultimate objects of thought, which correspond to the basic concepts. Truth and falsity are a function of the synthesis of concepts in judgement (cf. *Int.* 16a9ff.).

however, from the mental image of Socrates other than in that the concept 'Socrates' is an indivisible and simple motion of the soul-activity and is not an object of contemplation, whereas imagination manifests itself as an affection of thought, being a kind of working-out and accomplishment[18] of it. I am not saying that discursive intellection *is* imagination[19] (for there are intellections of incorporeal entities, but there
25 are not mental images of them), but rather that imagination underlies intellection. So then, predications are not the entities themselves, but utterances signifying concepts and things. For when they say, ' "Animal" is predicated of "man", ' they are saying that the utterance signifying *animal*, which is the word 'animal', is predicated of the concept signified by the
30 term 'man', and of the thing which is its subject; for to be predicated is a property of significant expressions, which signify concepts and things.

[4. How is one to identify the *Categories* and not be led astray by the difficulties raised by the Stoics?][20]

11,1 SELEUCUS: But how are we going to recognise, in the case of utterances which we are faced with, whether they fall under the categories or not? Give us some criterion, so that we may be able to mark off those things that do not fall under the division of the categories.

DEXIPPUS: I say, then, that the whole genre of 'the
5 significant'[21] must be postulated first as subject matter and set at the beginning of any such division; for it is not possible without the postulation of such a class of terms to predicate anything of anything else. So if we take an element of speech which is non-significant in itself, such as '*blityri*', or if

[18] The rare word *dianusis* used here is attested for Iamblichus at *De Myst.* 4.3, 184, and is also found, in the same combination as here, with *diexodos*, in Simpl. 309,11.

[19] cf. Aristotle's discussion of the relation of these two in *DA* 3.7, which Dexippus must have in mind.

[20] The Stoics mentioned here are almost certainly Athenodorus and Cornutus, who thought the *Categories* was a bad treatment of the parts of speech; cf. Porph. 59,10ff.; Simpl. 18,26ff.

[21] *Sêmantikon,* the opposite being *asêmon*, just below, while *sussêmantikon* I render 'co-significant' (cf. 32,17ff. below, where the term is explained). Plotinus is the first attested user of this latter term, in his criticism of the *Categories* (*Enn.* 6.1.5,4), though it is hardly likely that he invented it. He also uses it in a different sense from that used here, of a categorial term capable of signifying two categories at once.

something is significant by reference to something else, as in the case of so-called 'pronouns',[22] which indicate something by the employment of an indefinite part of speech, as for instance the word 'that one', referring to a particular person, or if a 10 term is 'co-significant' with something else, as is the case with articles and conjunctions, in no way would it be proper to include these among the predicates. Again, along with its relation to a particular thing and its distinction from things, the purpose of a category is properly to receive form,[23] since a significant utterance possesses language's primary function, 15 according to which we strive to reveal how things are related to one another, as for instance the term 'man' has its relationship to Substance, 'whiteness' relates to Quality, and so on. So if an utterance (*lexis*) is so structured as not to have the primary signification of language (*logos*), but rather so as to take as its defining characteristic its relation to its own parts, as for instance 'from Zeus', 'from home', 'most excellently', 'most correctly', 'most wisely', 'most poetically',[24] 20 or so as to take its meaning from the relation of the objects of thought to each other, as for instance in the case of the consequence of an hypothetical proposition such as 'If it is day', and the alternative contradiction of the disjunctive proposition, 'Either it is day ...', none of these would yet be an utterance suitable for predication; for all such uses of language (*logos*) are utterly alien to the primary relationship 25 of predicates to things.

Again, if there be some non-verbalised (*arrhêtos*) movement of the thought, such as is evidenced by groans and roars, or some inarticulate noise such as tut-tutting, or some indiscernible sound such as gibbering, or if some name signifies no reality (*pragma*), in no way is any of these a predication; for to no degree do they preserve the characteristic essence of a predication. Furthermore, the 30

[22] *Epanaphorai*, used here in a grammatical sense. The meaning is not, however, clear to me. Dexippus himself seems to be using the term rather tentatively (Steven Strange suggests 'anaphoric expressions').

[23] *Eidopoieisthai*. The meaning seems to be that categorial expressions get their essential differentiae from what they designate. Cf. 16,8-9 below, and Porph. 58,24ff.

[24] In all cases these are single words, either forms in *-then* (e.g. *oikothen*), or superlative adverbial forms, such as *kallista*. The point of these examples seems to be that the addition of a suffix does not yield a term which signifies a distinct categorial item.

highest distinctions of genera, or, if one prefers so to term them, of most generic expressions (*lexeis*), have nothing in common with each other, so they will have no further independent predicate of a substantial nature, either real or linguistic, superior to them. Following this principle of distinction, those who are seeking an accurate understanding of the categories should neither declare Being to be common to 12,1 all of them (otherwise there will no longer be ten, but all will be subsumed into one genus), nor should motion be predicated (*katêgorein*) as something common to acting and being acted upon; for then there will no longer be these two genera, but both will become one, to wit, Motion.

Again, it is the quality of simplicity and non-compositeness, 5 whether one sees it as inhering in the genera of being,[25] or in the most generic significant utterances, or in both, that everywhere constitutes the most proper defining characteristic of the categories. Therefore neither composite expressions, such as 'Dion is walking', nor quasi-composite ones, such as 'he scraper-manufactures',[26] nor expressions involving syncope or apocope, nor invented words, nor derivatives 10 (*pareskhêmatismena*), nor adjectives, nor words proper to poetry or rhetoric, are in any way relevant to the categories; for it is the business of another, secondary discipline to study such phenomena, namely philology. It is the role of the category to identify the primary signification (*sêmasia*) of utterances and the primary essential common features of entities, which are to be found in the highest genera, but not 15 the secondary common concepts (*koinai epinoiai*) of them which are manifested accidentally. Derivative forms (*skhêmatismoi*) and secondary concepts, which can be extended *ad infinitum*, are of no use for the attainment of knowledge (*epistêmê*), while generically signifying utterances or groupings of things under the primary genera, by applying limit to

[25] A reference, presumably, to the *genê tou ontos* of the *Sophist*, which since Plotinus (at least) had come to be regarded as the proper 'categories' of the invisible world.

[26] Reading *xustropoiei* for meaningless *xustropôn* of mss. What is required is a compound verb, to make a *huposuntheton* expression. The Latin translator Felicianus gives *multifidus*, which has no apparent relation to the Greek, but preserves the idea correctly. Busse relevantly adduces Simpl. 71,28, where, as examples of *huposuntheta*, we find *lithobolei*, *pseudodoxei*, *boukolei*.

the unlimited, instil accurate knowledge of language at the same time as of things (*onta*).

For this reason we must make use of these guidelines as being those which are, reasonably, particularly favoured by serious students of language and substance (*ousia*). Therefore neither will a figurative nor metaphorical expression such as 'upset the apple-cart' (*anekhaitisen*)[27] or 'lowest foot of Ida' (*Iliad* 2.824), nor yet modal expressions such as 'necessarily' or 'possibly' (*endekhesthai*) or 'actually' (*huparkhein*), nor quantifiers (*prosdiorismoi*) such as 'all' or 'no' or 'some' or 'some ... not', fall under the categories; for none of these has a distinct objective reality specifically corresponding to it, so that one might reasonably dismiss them as not conforming to the previously mentioned purpose[28] (*skopos*) of the *Categories*.[29] I think, then, that it is that aspect of language that relates to truth and that in which truth and falsity are properly involved that is primarily suitable to the categories. So then, it is such portions of significant speech as, being taken together to form a simple predicative statement, give proper value to the terms out of which they are constructed, that would be the subject matter of the categories. 20 25 30

But let us, for the sake of clarity, continue our exposition in the form of question and answer; for, while there are many possible areas of enquiry on these subjects, it is better to concentrate on the topic before us. Ask on, then, if you will. 13,1

SELEUCUS: To what category does Being belong?

DEXIPPUS: To none.

SELEUCUS: Why? 5

DEXIPPUS: Because it is homonymous throughout all ten.

SELEUCUS: Well then, to what category does the One belong?

DEXIPPUS: Again, to none.

SELEUCUS: Why?

DEXIPPUS: Because this also is said in ten ways, like Being. 10

SELEUCUS: The opposite of One is Plurality (*plêthos*). To what category, then, does Plurality belong?

[27] Literally, 'tossed his rider'. Certainly intended as a well-known reference, probably to Dem. *Olynth.* 2,9, where it refers to Philip's activities.

[28] Reading *proeirêmenôi* for *proeirêmenôn* and excising *tôn nun*, as Busse suggests.

[29] The book title seems intended here.

DEXIPPUS: To none.

SELEUCUS: Why?

DEXIPPUS: Because Plurality too is said in ten ways – in the area of Substance, as in 'men'; in that of Quantity, as
15 'three-cubits-long'; in that of Quality, as 'white(s)'; in that of Relation, as 'fathers'; and so on.

SELEUCUS: Well, how about Not-Being; to what category does it belong?

DEXIPPUS: Not-Being is said in two ways, the one signifying the removal[30] of every aspect of reality (*phusis*), which is absolute not-being, neither substance nor quality nor any of the other categories; or one may have relative not-being, as for instance *not* quality, *not* substance, but something else, such as relation.

20 SELEUCUS: To which category, then, does either of these senses of not-being belong?

DEXIPPUS: Absolute not-being belongs to no category, for it is entirely without substance (*anuparkton*). But relative not-being can actually be taken in ten ways – in one way as not-substance, in another as not-quality, and so on.

SELEUCUS: And 'is' – to what category does it belong?

25 DEXIPPUS: This also belongs to no category.

SELEUCUS: Why?

DEXIPPUS: Because it is homonymous throughout all ten.

SELEUCUS: And 'same' and 'other' – to what category do they belong?

DEXIPPUS: Each of these too can be taken ten ways; for a
30 substance can be 'the same' as a substance, or a quality as a quality, or a quantity as a quantity, and so on. But if you go across (*enallax*) categories, there will be more senses of it – Substance in relation to Quantity, and Substance in relation to Relation, and so on, making another ten senses of it, and if you start from Quality, there will be another nine (Quality in relation to Quantity and Quality in relation to Relation, etc.), and starting from Quantity there will be another eight, and from Relation another seven, and so on analogously with the
35 rest. And similarly 'other' will be said in ten ways, either as
14,1 Substance in relation to Substance, or Quality to Quality, or

[30] Or, reading *arnêsis*, 'denial'.

Quantity to Quantity, and so on. And by crossing categories you will get an analogous further plurality.

SELEUCUS: Each of the categories is something (*ti*)?[31]

DEXIPPUS: Yes.

SELEUCUS: Well then, how is each of them something? 5

DEXIPPUS: In ten ways, firstly as Substance is something, then as Quantity is, and so on for the others.

SELEUCUS: And the definition (*horos*), does it indicate the essence[32] of a 'thing'?

DEXIPPUS: Yes indeed.

SELEUCUS: In what way? 10

DEXIPPUS: The definition indicates the, as it were, proper characteristic (*idiotês*).

SELEUCUS: What do you mean by 'proper characteristic'?

DEXIPPUS: This too has ten meanings, the one as proper characteristic of a substance, another as of a quantity, another as of a relation, another as of a quality, and so on.

SELEUCUS: Is 'definition' univocal, then? 15

DEXIPPUS: No. This is said in ten ways, even as many as being.

SELEUCUS: Why?

DEXIPPUS: If the definition provides the differentiating features of each thing, and Being is said in ten ways, then 'proper characteristic' is said in ten ways.

SELEUCUS: A definition is made up of genus and proprium (*idion*). What is meant by genus and proprium? 20

DEXIPPUS: These terms also are said in ten ways. For there are as many genera as there are categories, and there are as many sets of propria as there are categories.

SELEUCUS: Does genus differ from category?

DEXIPPUS: In one way, yes, in another no.

SELEUCUS: How yes, and how no? 25

DEXIPPUS: In so far as they relate to a subject, no (for the types of subject are also ten), but inasmuch as categories are utterances and genera natures (*phuseis*), they are different. (And similarly with the differentia; for there is a differentia

[31] This query may have something to do with the fact that the supreme Stoic category is *ti*; cf. *SVF* 2.329, 332-3, but it may just reproduce the basic question that Socrates commonly asks, viz. 'Is *x* something?' = 'Is there such a thing as *x*?'

[32] *to ti esti*, lit. 'the what-it-is'.

for each genus, and there are ten senses of genus. And since the division of a genus is into species, the split would be ten 30 ways, even as many ways as the concept 'genus' is split. And the concept 'species' is split into an equal number of parts.)[33]

SELEUCUS: How is it then, if 'genus' is ambiguous (*pollakhôs legetai*), Aristotle gives as a definition of it, 'that which is predicated in answer to the question "what is it?" of several things which differ in species'?[34]

35 DEXIPPUS: Because it is quite possible for a definition of 15,1 ambiguous things to be itself ambiguous; for even as 'genus' is said in ten ways, so that which is predicated as an answer to the question 'what is it?' is said in ten ways.

SELEUCUS: But if it is the fact that Aristotle has said that 'same' is said in three ways (for a thing can be the same in either genus or species or number),[35] how is it said that 'same' is said in ten ways?

5 DEXIPPUS: Because being the same in genus is said in ten ways. For instance, we have things the same in substance, such as 'man', 'horse' (for they have the same genus, Animal, and Animal is a substance); and in quantity, such as line, surface (for they are the same in size),[36] and in the area of relation, as for instance double <and triple, for they are 'the same', as being multiples>;[37] and in quality, such as 'white', 'black' (for each of them is a colour), and the same in the case of the rest. And similarly, the same in species is said in ten 10 ways – either in substance or in any of the rest, and the same numerically likewise. And in the same way 'other' is said in ten ways. And seeing as he has made the triple division by genus, species and number, I declare that the generically other, the specifically other and the numerically other all have ten meanings, and examples of this are obvious.

SELEUCUS: 'Two ways', 'three ways', and 'in two', 'in three' 15 and 'in four' – what category will we put them under?

DEXIPPUS: We have already stated repeatedly that, of

[33] Presumably this is the meaning of *kai to eidos eis isa temnetai*; the whole passage within brackets does not fit well here. Busse suggests that it is the marginal note of a scholiast, and goes best with the content of ll. 20-2 above.

[34] *Top.* 1.4, 102a31.

[35] *Top.* 7.1, 152b31ff.

[36] Excising *posa* after *esti*, as Busse suggests.

[37] This passage is corrupt. I give a possible supplement, following Felicianus' Latin version.

utterances (*phônai*), some are possessed of primary signification (*sêmasia*), while others are to be referred to a secondary use of the language (*logos*), which latter the theory of categories makes no claim to deal with.[38] For this reason if one were to refer these to the primary concepts (*noêseis*), starting out from which the intellect (*dianoia*) generates these motions, one would declare them to belong to the category of Quantity; but if one divides these off from the primary concepts, which are directly related to things, one should not ask to which class of the enumerated categories they might belong. 20

SELEUCUS: Turning to 'whole' and 'part' – into what category will we place them?

DEXIPPUS: These too belong to the secondary use (*khreia*) of language. 'Hand' and 'head', for instance, are primary names (*onomata*), but that hand and head are parts of a whole, and are named as parts, is an instance of the secondary level of linguistic reference.[39] 'Socrates', for instance, is a name of a substance, but the statement that 'Socrates' is a name is a second-level employment (*thesis*) of the name on top of the primary one. Such, then, is the case with whole and part, and for this reason we may be excused from referring them to the categories. But if we were to do so, we could class them as relatives; for the part is part *of* a whole and the whole is a whole *of* parts. 25 30

SELEUCUS: 'In general' and 'once and for all' and 'wholly'[40] – to what category shall we refer *them*?

DEXIPPUS: These are in dependence upon[41] the categories; for they derive their reference from participation (*kata* 16,1

[38] The backward reference is to 11,12ff. This contrast between primary and secondary level or 'use' (*khreia*) of language is interesting. The primary level names things, the secondary comprises, it seems, words which express relations between concepts. This corresponds to a distinction made by Porphyry, 58,1-3 between the *prôtê* and *deutera thesis tôn onomatôn*, the 'secondary' type of use being the use of words to talk about *language*, not things. Cf. the useful discussion and collection of texts in Hoffmann, 1987, 78-91.

[39] If this is a proper rendering of *deutera kinêsis lektikês menuseôs* (*menuseôs* itself being a correction in ms. A for *kinêseôs*). This is presumably just an elaborate variant on *deutera tôn logôn khreia*.

[40] *Katholon*, *kathapax*, *ardên*, all modal expressions.

[41] I prefer to read *exêrtêtai* with A, to the *exêirêtai* of CMR, which Busse chooses. That would mean 'transcend the categories', but that does not seem to be the meaning required. These modal expressions are to be seen, presumably, as dependent upon their corresponding simple terms.

metokhên) in certain primary and archetypal categories.

SELEUCUS: The same and the other and the proper (*idion*) and the incompatible (*allotrion*) and the common, and indefiniteness and ambiguity and synonymity – where are we 5 to rank them?

DEXIPPUS: It has been said already that each of these is said of all the categories; but since they relate to names, they fall outside the scope of the categories. For we have already said repeatedly that the categories derive their existence (*huphistantai*) from their signifying (*sêmantikon*) the distinguishing characteristics (*diaphorai*) of existent things; 10 wherefore we all the more think it correct to see individuals (*atoma*) and things that can be pointed to (*ekhonta deixin*) as most properly in the categories and as able to satisfy to a greater degree their stated aim. So then, having sorted out these questions, we may suitably turn to the study of the categories at the highest level, applying them to our findings so far.

[5. Why, when he announces his intention of discussing the categories, does he not start with them, but with the subject of homonyms and synonyms?]

SELEUCUS: Why on earth, then,[42] having put it forward as 15 his purpose to discuss the categories, does he not start by talking about them, but talks rather about homonyms and synonyms and paronyms, although in his other writings he is accustomed always to stick to his main subject, and to deal with nothing beyond what pertains to his stated theme?

DEXIPPUS: We shall say that names are not adequately matched to the multiplicity of things, nor do things have an equal number of significatory locutions corresponding to them, 20 neither individually, nor in common, generically, but in some cases there is a superfluity, and in other cases an inadequacy, of the one in respect of the other; for indeed there are often found a plurality of names for the one thing, and one name for

[42] We now return to subject matter which is paralleled fairly closely in Simplicius (21,1ff.), and less closely in Porphyry (60,1ff.), who provides a different answer to this question. Dexippus is, however, not particularly close to Iamblichus either, as quoted by Simpl. 22,1-9.

a plurality of things, formed according to no common or identical principle,[43] so that one must consider in regard to the proposed simple and incomposite utterance, into what category it should be placed. It is for this reason that he deals with homonyms, synonyms and paronyms, by way of 25 teaching us to observe from every angle lest something such be applicable to the utterances we are dealing with, so that he first lays down what should properly be established first by those who are proposing to acquire accurate knowledge of utterances which fall under categories. For even as in the case of the other sciences we establish certain principles in advance, without which it would be difficult, if not impossible, to have knowledge of the subject matter, so here too, prior to 30 the division into categories, we establish in advance some principles which will contribute to the clarification of what follows.

[**6**. Why did Archytas in his Generic Discourse[44] not discuss homonyms and synonyms and paronyms?]

SELEUCUS: Well, should not Archytas then, prior to his discussion of generic utterances, which we call categories, have discussed these things?

DEXIPPUS: Perhaps such distinctions are not in accord with 17,1 Pythagorean principles (*nous*); for since they lay down that names are attached to things by nature (*phusei*), they deny all anomaly in language.

[**7**. Why did Aristotle not prefix everything useful, but produced other things after his presentation of the categories?]

SELEUCUS: Well, let us leave that aside, since this sort of enquiry would be more properly directed to those who profess 5 Pythagorean doctrines; but if it was his decision to prefix

[43] i.e. not applied to the same common nature in each thing (as 'animal' is in Plato and Fido).

[44] This title for Archytas' work is the same as that used by Boethius in his *Commentary on the Categories* (I, 162A (Migne)). Simpl. calls it *Peri tou pantos* (2,17) or *Peri tôn katholon logôn* (e.g. 13,23; 40,5). He also tells us that Iamblichus was the first to make use of Archytas in the exegesis of the *Categories* (2,20ff.).

useful principles, why did he wait till after the presentation of the categories to bring up the discussion of contraries and contradictories (*antikeimena*) and motions and the rest?

DEXIPPUS: Because, my dear fellow, in the case of the other things of which we have no conception (*ennoia*), it is reasonable that we should learn of them in advance from
10 those who have expertise in the area, and homonyms and synonyms and paronyms are of this sort; for we do not have any preconceived notion (*prolêpsis*) of these, and it was therefore surely necessary that we should be introduced to those concepts by someone who used the words accurately. But in the case of those things of which we have a natural conception (*epibolê*), even if not an articulated one, one needs to make use of common sense (*koinê ennoia*) in order that the
15 continuity of the exposition not be interrupted, and after that to make an accurate definition of each, distinguishing the things (*pragmata*) themselves, and not employing physical theory[45] to examine them. Straight away, then, he brings the instruction round to genera of a sort, in treating of contraries and contradictories and the rest, and further on the subject of motion and the simultaneous, employing more crude (*pakhuterai*) concepts, whereas he gives a more accurate
20 account in the *Physics*.[46]

[8. Why does he speak about homonyms?]

SELEUCUS: What is the reason for his mentioning first homonyms, rather than synonyms, although homonyms have in common only one thing, the name, whereas synonyms have both, I mean the name and the definition (*logos*)?

25 DEXIPPUS:[47] I declare that the reason for this should be referred to the categories themselves; for since these happen to be homonyms, it was reasonable that he should put first that which is primarily relevant to them all in common. For each of them individually is categorised synonymously by the

[45] The *Categories* is not supposed to be a work of physics, but of logic, and hence cannot employ any *phusikê theôria*.

[46] cf. *Physics* 3.1, 200b12ff.

[47] This reply of Dexippus' is paralleled very closely by Simplicius 24,1-5, a passage which is explicitly attributed to Iamblichus. Cf. Porph. 61,10ff., where the point is made that *Being* is a homonym.

species subject to it according to its own proper genus, whereas all alike are termed categories homonymously.

[**9**. Why did he not rather speak about homonymy?]

SELEUCUS:[48] But if it was his purpose to discuss utterances both homonymous and otherwise, why does he say nothing about homonymy, but teaches us instead about homonyms, which are conceived of as having their doubleness of signification in virtue of being relative to things? 30 18,1

DEXIPPUS: Because homonymy also is discerned, if we proceed correctly, on the basis of the fact that *things* are different. So he starts out by teaching us those things from which it primarily takes its constitution and which are most properly its causes and in which it has its being, when we cognise which, the relevant utterance is at the same time apprehended as homonymous; for when these things are multiple, straightway the names also are shown as having multiple meanings. So it is reasonable that he should prefix those things of which the discovery is easiest and most effectual. And these are things which are the effective producers of homonymy, so that it is in accordance with the distinctions between things that he presents homonyms as following from these, and so gives his account of homonymy also. 5 10

[**10**. Why does he add 'only' in saying 'have only a name in common'?]

SELEUCUS:[49] But if the term 'only' has two senses, either the sense of uniqueness, as in 'there is only (one) sun',[50] or by way of contrast to other alternatives, as in 'having only a *khitôn* (tunic)', when he says homonyms are what have only a name in common, and the definition of being corresponding to the 15

[48] This *aporia* is in Simplicius (24,6-9), but the reply is not verbally close to Simplicius. The point of the question is that, if the *Categories* are about *lexeis* rather than *pragmata*, then it would be more suitable to talk about homonymy rather than homonyms, since the latter are attributes of things. Cf. Porph. 61,13ff., where the terms used are *phônai* and *pragmata*.

[49] This section corresponds loosely to Simplicius 26,3-10, with enough verbal correspondences to indicate that the same source is being drawn on.

[50] cf. Porph. 62,5-10: 'there is only (one) cosmos'.

name is different, either the 'only' is meant by him to signify a contrast with definition (*logos*), or it is not. But if it signifies that, then the rest of the phrase is superfluous, whereas if it doesn't, then the 'only' is superfluous.

DEXIPPUS: My reply is that these additions are for the sake
20 of clarity, which should be a prime consideration for speakers. For here the ambiguity of 'only' is resolved by what follows; for as he who says 'this man has only a *khitôn*' makes his statement clearer by adding 'and he does not have a tunic', so also he who adds to 'only' 'and the definition of being which corresponds to the name is different' is exhibiting a proper concern for clarity.

[**11**. How comes it that, whereas many things are common to homonyms, he uses 'common' in the case of the name?]

25 SELEUCUS: But I would say that there is not only the name in common in the case of homonyms but many other things as well (for instance, the two Ajaxes have in common not only the name, but being Greek and generals and friends). How, then, is it that he says 'have only a name in common', when there are many things common to homonyms?

30 DEXIPPUS: Because we are not investigating what is common to the Ajaxes either in terms of linguistic description or kind, but only in so far as they are homonymous with one another, so that the addition of 'only' is essential; for the homonym is homonymous only in respect of the name.

[**12**. That the name is common to homonyms neither in the sense of the indivisibly common nor in the sense of the divisibly common.]

SELEUCUS: But if 'common' is an ambiguous term[51] – for you can have something in common like a horse or a slave which is
35 indivisible into parts, and again something that is divisible,
19,1 and this in various ways, either what can be appropriated for the use of each individual, but after its use returned to

[51] cf. Porph. 62,19-28.

common possession, as for instance, a bath-house or a theatre, and also that which after its apportionment becomes the private property of each individual, as do the spoils of the enemy for soldiers – if, then, none of these modes of being common can be applied to homonyms, then their existence is demolished; for homonyms partake of the name neither as indivisible, as are a horse and a man, which come into (common) use by turns (for there is nothing stopping many simultaneously employing the same name), nor as divided (for participation in the name is not apportioned by syllables). But if it is common neither as an indivisible nor as a divided entity, how is it possible for a name to be 'common'?

DEXIPPUS: Whoever raised this difficulty ignores the fact that, of the indivisible, one type is said to be taken for individual use consecutively, e.g. the horse and the lyre, while that which can be availed of simultaneously and undividedly by many is left as indivisible proper, and it is this latter that is the mode of the commonness of the name; for it is precisely by virtue of its not being divided that it is present to all entities that partake in it.

[**13.** A further problem relative to 'common'.]

SELEUCUS: But if the name fully (*aprosdeôs*) signifies that of which it is a name (for if it needed some other element in order to signify, it would not be a name), then that which is common to many does not signify one thing. For it is unclear[52] which of the many it intends to signify; e.g. he who says 'dog' makes plain that what is signified is not an ox or a man, but does not signify what kind of 'dog' he means.[53] But the name is that which is significant, so that it would not be possible for a name to be common to more than one thing; if this is the case, there will be no such thing as a homonym; for the theory of homonyms postulates that a plurality of things are signified by some common name. But if the name needs an addition in order to signify that one out of the many which it intends,

[52] This sentence occurs virtually verbatim in Simpl. 26,22-4, the *aporia* being attributed to Nicostratus. Since Dexippus reads as if he is summarising the version presented in Simplicius, the probability is that the passage in Simplicius is taken from Iamblichus.

[53] i.e. terrestrial dog, dogfish, or dogstar.

then once again the definition of the homonym is overturned; for it no longer remains common, when it becomes particular to one item by the addition of a differentia, as when we add to 'dog' the qualification 'terrestrial' or 'aquatic', so that not only does the name not remain common, but it actually ceases to be
30 a name at all, having taken on a differentia.

DEXIPPUS: While rejecting the captiousness of the objection, we must not refuse a solution of it. A wrong sense has been
20,1 given to the 'commonness' of the name; for Aristotle did not intend 'common' in the sense in which that is applicable to the lyre or the horse, but rather in the sense in which the voice in the theatre is 'common', in that it, while being one and the same, extends to the hearing of different people; for in the present case also the name, while being one and the same, is related to various different things.[54]

[**14**. Yet a further problem relative to 'common'.]

5 SELEUCUS: But if someone were to say that everything subject to sense-perception is changeable (*treptos*), and everything changeable is subject to alteration and everything subject to alteration never stays the same, but comes to be different at different times – whether one takes the alteration to be in multiple respects or in just one – and that which alters cannot be 'common' (for the common should persist in one state; if it does not, then what accrues to each entity becomes
10 individual to each); but a name is sense-perceptible – for sound (*phônê*) is something sense-perceptible; so on this argument as well there would be no such thing as a common name; and if this is so, then no homonym. I realise myself how sophistic such an objection is, but I would still like to be given a solution to it.

DEXIPPUS: I say that they are ignorant of the fact that 'common' is not being used here in the sense of what is

[54] This solution (which is that approved by Porphyry 62,30ff.) is actually rejected by Simplicius (27,12-15), who makes the pertinent point that a voice is heard the same by all, whereas the homonym is precisely to be understood differently in its various applications. He then goes on to present an argument by 'some authorities' that not every name needs to be significant. If Simplicius is dependent on Iamblichus here, we would have an interesting case of Dexippus following Porphyry rather than Iamblichus, but we cannot assume that Simplicius is not being original.

divisible among a multiplicity, but as being the same as a whole for each, and not one thing for this one and another for that, so 15 that the commonness of homonymity would be an aggregate of individual[55] forms. In fact, on the same terms that the individual name of each thing perseveres, so does the common name also persevere. And further, even if 'changeable' is applied to sound as to a body, nevertheless this must not be taken as applying to the actual name; for this must remain unchanged.

[**15**. That it was superfluous of him to add the phrase 'the account (*logos*) of being which corresponds to the name'.]

SELEUCUS: Since they do not only complain about this, but 20 also bring the charge of superfluousness against his description, this too needs to be examined. For they say that it would have been enough to say 'When things have only a name in common, *and the accounts are different*, they are called homonyms.'

DEXIPPUS: But if 'the account of being which corresponds to the name' had not been added, it would have been possible to 25 show that the same things were homonyms and synonyms, as for instance the Ajaxes, for if the definition had not been specified as 'corresponding to the name', one and the same definition could have applied to them as men,[56] and so they would have turned out to be synonymous.[57]

[**16**. Why did he not say 'definition of being', but 'account'?]

SELEUCUS: But why ever did he not say 'the *definition* (*horos*) of being is the same', but rather 'the *account* (*logos*) of being'?

DEXIPPUS: Because not all things are demonstrated by 30 definitions, but also by descriptions (*hupographai*), and for

[55] Reading *idiôn* for *idian* of mss. I cannot attach a satisfactory meaning to the text as it stands.

[56] i.e. 'rational mortal animal'.

[57] This explanation is given in a fuller form by Porph. 64,9-21, and Simpl. 29,2-12.

this reason he was accurate in using 'account' of being instead of 'definition'.

[**17**. That homonyms can be synonyms.]

SELEUCUS: Since some people insist on trying to show that homonyms are synonyms, please clarify this matter also. For they say that if those things are synonyms of which the name and the account corresponding to the name are the same, and 21,1 homonyms both have the same name (for they are called homonyms) and admit the 'account' (*logos*) of homonym, then homonyms will be synonyms; but synonyms also are *a fortiori* synonyms; therefore all of them are synonyms.[58]

DEXIPPUS: Ideally one should not even attempt a reply to such captious arguments, except that we should counter 5 them, lest they seem to have some substance. There is no problem about things being according to one designation homonyms and according to another synonyms; so here too, all homonyms are given the name 'homonym', as for example the Ajaxes; for each of them is Ajax and each of them is a homonym, and *qua* Ajaxes, having no common reality (*pragma*) applying to both of them, they are so-called 10 homonymously, while *qua* homonyms they do have a 'common account', and so become synonyms.[59]

[58] This *aporia* is presented by Simpl. at 30,16-22, who identified it as emanating from Nicostratus (and clarified further by Atticus).

[59] This *lusis* (solution) is given by Simpl. (30,23-30) as that of Porphyry (it does not occur in his small commentary). Dexippus may have adopted this directly, but more probably through Iamblichus. There is an interesting variation in the argument between Porphyry and Dexippus (if we can trust Simplicius, and the manuscript tradition!) – Porphyry says: 'There is nothing to prevent the same things under different headings (*prosêgorias*) being both homonyms and synonyms, e.g. the Ajaxes, who may be called both "Ajax" and "man" [n.b. this phrase is omitted in one MS and may be a gloss], in so far as they are *men*, are synonyms, whereas in so far as they are "Ajax", they are homonyms. So then, the *homonymous* Ajaxes, in so far as they are homonyms, are synonyms, and in so far as they are "Ajax" homonyms.' The idea of being synonymous as *men* (and *anthrôpoi* is also read for *homônumoi* in one manuscript in l. 27) is one not introduced by Dexippus, and is not necessary for answering Nicostratus' *aporia*. The main point is that, as *homonyms*, the Ajaxes are synonymous, while as 'Ajax' they are homonymous, so that Aristotle's distinction is shown to be perfectly valid.

[18. How comes it that whereas homonyms may be seen in all the categories, the account of them comprises homonyms only in the category of Substance?]

SELEUCUS: But how are we to reply to those who claim that the description of homonyms is inadequate? For they say that if homonymy is to be found not only in the category of Substance, but in Quality and Quantity and the other classes as well (for after all one can employ homonyms in all those categories), how comes it that he used the phrase in his account of homonym 'the account *of being* (substance)[60] which corresponds to the name'? This seems to be an inadequate expression, if it is true that, whereas homonymity occurs in all the categories, this account (*logos*) only covers homonyms in the category of Substance.[61] 15

DEXIPPUS: First of all, the expression 'account *of being*' does not occur in all manuscripts, as both Boethus and Andronicus testify.[62] But even if we accept the form 'the account of being is different', because it is found thus in the majority of manuscripts, nevertheless the form of expression is correct. For if the word 'account' is a homonym (for both syllogism and inductive argument are a form of account, but the account which indicates being is different from both of these), he did well to add 'the account *of being* is different'. Furthermore, since Substance is not the only thing said to 'be', but accidents also are said to 'be', he spoke reasonably in saying 'account of being', making reference not only to Substance, but to the other genera as well, in so far as they all enjoy *being*; for the account that reveals the existence (*huparxis*) of each of the accidents would rightly be termed the account of their being. So then, we may take 'account of being' as a proper description of the essence (*hupostasis*) of homonyms. 20 25

[60] This whole objection rides on the Greek word *ousia* being used both for the category of substance and for 'being' in general. Since I am unwilling to give up the normal appellation for the first category, this will have to be understood by the reader.

[61] This *aporia* is attributed to Nicostratus by Simpl. 29,24-8.

[62] This information goes back to Porphyry, whose *lusis* this is (Simpl. 29,28-30,15) – no trace, however, in Porph.'s short commentary. Once again, however, this does not preclude the intermediacy of Iamblichus. The point about 'being' applying just as well to accidents as to substance (which is surely the correct reply to the problem) is not found in Simpl.'s account of Porph., and so may either be attributed to Iamblichus, or (improbably) to Dexippus himself.

[**19**. Why did he put synonyms after homonyms?]

30 SELEUCUS: Well then, if synonyms partake of both the same account and name, why ever did he omit to discuss these first, 22,1 but rather made mention of homonyms?[63] And further, if he is going to give an account of the highest genera, and it is a property of (*huparkhei*) genera to be predicated synonymously, and not homonymously, of those things that fall under them, then synonymy, but not homonymy, is proper to them. But in all cases what is a property of something is prior to 5 what is not a property of it, so that he should first have explained synonyms, and only then homonyms.

DEXIPPUS: Because Being is not according to Aristotle a synonym, as is the view of others, but rather a homonym. If, then, Being itself is an homonym, the doctrine of the homonym takes precedence, and the demonstration of what is not (homonymous?),[64] for one must first define homonymy, 10 and then, having thus enumerated and defined the genera individually, go on to demonstrate how they are predicated synonymously of those things falling under them.[65]

[**20**. That the 'in-combination/without-combination' division is erroneous.]

SELEUCUS: There are some who criticise the division into things 'in combination' and things 'without combination'. For it is not the case that all utterances are made either 'without combination' or 'with combination'; for the expression, 'a man 15 is walking' is not 'with combination' (for it does not have a combinatory conjunction),[66] and on the other hand 'man' or 'ox' is not 'without combination'; for the very names possess a certain combination of syllables and letters.[67]

DEXIPPUS: We say that those who call 'combination' only an expression involving a combinatory conjunction are following

[63] cf. Porph. 61,6ff., who provides just the same solution.
[64] The text may be defective here; the reference of *tou mê huparkhontos* is not clear.
[65] Dexippus pays no further attention to synonyms, and none to paronyms, unlike Porphyry, who does (68-70).
[66] Such as 'and'. Cf. the fuller discussion in Simpl. 42,9-43,25, where, however, this *aporia* is not raised as such. Cf. also Porph. 71,4ff.
[67] *anthrôpos*, unlike 'man', has three syllables.

Stoic usage,[68] whereas Aristotle is earlier than the Stoics and is employing the usage of an older generation, who termed 'combination' the putting together of more than one part of speech. And as for saying that 'man' or 'ox' is 'in-combination', because there is in them a composite of syllables and letters, that also is not accurate; for it is not the composite of letters, but rather that of parts of speech, that the ancients call 'combination'. 20 25

[**21**. That the description of what is 'in a subject' can be held to include a body's being in time and place as in a substratum.]

SELEUCUS: But if indeed 'in a subject' is defined as 'being in something not as a part' (for it is impossible for that to be apart from that which it is in), and it is impossible for body to be separated from place, then body will be in place as in a subject (*hupokeimenon*).[69] And similarly also for what is in time; for neither are they parts of it, nor can they exist without time; for even if one were to say that they are separated from particular place and time, yet they are not separated from time and place without qualification. 30

DEXIPPUS: In reply to this also we will say that none of the common natures (*koina*) is either a 'this' or a 'something', and so that which is in generic place or time will not be 'in something'. Further, that which is essentially in the sphere of time is that which is moved or is subject to change, for it travels and changes along with time (if there were something sempiternal,[70] it would only accidentally be subject to time, as being unmoving and unchanging; for it is said to exist while time exists). If we accept this, then, if what moves, in so far as it moves, changes with time, it would not be 'in something' in the strict sense, nor yet 'in time'. For if, when moved, it were not in place, it would not be 'in time' either; for that is the 23,1 5

[68] The term *sumplektikos sundesmos* is indeed first attested for Chrysippus (*SVF* 2.207).

[69] Which may also be translated 'substratum', of course. On this problem, cf. Porph. 79,12ff.

[70] I take *aidios* (as opposed to *aiônios)* to have this technical meaning. The point about sempiternal things existing along with time is an allusion to Ar. *Phys*. 4.12, 221a9ff., 221b2ff.

same situation as if it were moved spatially. Then far more so it would not be in place, since both (place and time) are essentially in motion themselves – so it would have the same experience in respect of time as well; for neither of them holds
10 still, so that the one will not *be* in the other, neither in the particular nor in the generic. But if something must be said to be 'in time', it must be said in a more general sense, not in a defined stretch of time; for there will be no means of indicating either motion or time, since both are considered as having extension (*en platei*), and come into being sequentially[71] and according to the continuous generation of their parts; for one part is past, while another is yet to come. But if this be
15 the case, then being in place and time must be held to be different from 'being in a subject', as in the case of a colour.

[**22**. That what are in a subject are found to be parts of the substance.]

SELEUCUS: If those things that constitute (*sumplêroun*) the substance of each thing are parts (for this is the concept of a part, that the part is constitutive of the substance), and the substance of a body simply so-called is constituted by colour
20 and shape and quantity (for no body is without colour or shape or quantity), and the substance of this particular body, e.g. fire or snow, is constituted by cold or heat or whiteness or brightness, then either the aforementioned things will be parts of them and what are in a subject will be found to exist in it as parts, or they will not be parts and will not constitute the substance.[72]
25 DEXIPPUS: To this difficulty one ought to make the following reply, that 'subject' has two senses, both with the Stoics and with the older philosophers,[73] one being the so-called primary subject, i.e. qualityless (*apoios*) matter, which Aristotle calls

[71] If this be the sense of *kata diexodon*. A similar meaning occurs in Simpl. 118,21: *tôn kata diexodon kinoumenôn*.

[72] This is given by Simplicius (48,1-11), as an *aporia* of Lucius, and what follows in Dexippus is identified as the solution of Porphyry (48,11-33). The verbal equivalence is very close, but this still does not exclude Dexippus, at least, having derived Porphyry (and, of course, Lucius) directly from Iamblichus. It does not figure in Porphyry's short commentary. We may note that this very problem is the subject of Plotinus' *Enn.* 2, 6, 'On Substance' or 'On Quality'. Cf. also *Enn.* 6.3, 5.

[73] sc. the Peripatetics and the old Academy.

'potential (*dunamei*) body',[74] and the second type of subject is the qualified entity, either general or particular; for both the bronze and Socrates are subjects to those things that come to be 30 in them or are predicated of them. For 'subject' is regarded as being a relative term (for it is the subject *of* something), either without qualification, of those things that come to be in it and are predicated of it, or in a particular sense. Unqualifiedly, the subject for all accidents and predicates is prime matter, while for particular accidents and predicates the subject is, e.g. the 24,1 bronze or Socrates. So then, there being two sorts of subject, many of the things which come to be in something, while in relation to the primary subject they are 'in a subject', in relation to the secondary one, they turn out to be not *in* the subject but rather parts of it. Wherefore all colour and all quality and all shape are in prime Matter as a subject (for they 5 subsist in it not as being parts of it and as being unable to be separate from it), whereas on the other hand not every colour nor every shape attached to a secondary subject is *in* it, but only when they are not constitutive of the essence will they be *in* the subject. <For instance, white in the case of wool is in it as in a subject>,[75] while in the case of snow it is not in it as a subject, but rather it *is* the subject,[76] for the whiteness of snow is 10 neither said to be of another body as subject, nor *in* the body of the snow as a subject, but rather it completes the subject as a part. Aristotle, therefore, taking the secondary subject as the subject of his instruction, says that everything which does not inhere essentially is *in* something as in a subject, positing those things which constitute the essence as existents (*onta*), while those that do not do so he does not call parts, but attributes 15 (*sumptômata*) and accidents (*pathêmata*), on the ground that the qualities and quantities are attributes and accidents of the primary subject, while of the secondary subject they are parts, when they constitute its essence.

[74] Aristotle does not actually say this, though he implies it (e.g. *Metaph.* 9.8, 1050a15; 14.1, 1088b1, 14.4, 1092a3), but he was generally held in later times to have done so, as was Plato (Albinus *Did.* 163,6ff.; Apuleius *de Plat.* 192; Diels, *Dox. Gr.* 567, 16). It is worth noting that Porph. (quoted by Simpl. 48, 13) omits 'body' here.

[75] Added from the parallel text of Porphyry, ap. Simpl. 48,21-2.

[76] This is expressed more clearly by Porphyry (ap. Simpl. 48,23-4), as follows: 'but rather it completes the essence as a part, and *is* the subject, rather, as being the essence.' Dexippus (or Iamblichus?) chooses to expand this into an explanatory sentence.

[**23**. That the account given of what are in a subject is common also to what are not in a subject.]

SELEUCUS: But even if one were to grant that what are in a
20 subject are not parts, yet this is a strange fact, that in the definition of what is in a subject there can often be found both things that are 'in a species' and 'in a genus' and as wholes in parts; for the above-mentioned definition of what is in a subject is common to these as well. For neither is the species part of the genus (for even when the species is destroyed the genus can survive, as for instance Animal, if Man were
25 destroyed), nor could it survive without there being such a thing as 'animal',[77] for how could it be a species if there were no genus? But also the whole, although being in its parts, neither is a part of them (for it would be absurd for the whole to be a *part* of its parts), nor can it be apart from them. So the account has been shown to be common not only to what is in a subject, but also to things that are not in a subject.

30 DEXIPPUS: It is plain to everyone that the species is not a part in the way that a hand is of the whole body, but it is considered as a part. But Aristotle did not say 'what exists in something but is not a part of it', but 'not existing in it *as* a
25,1 part', so that he eliminated by this addition what is in something *as a species*. And the whole is not in any one of its own parts, but it is viewed as being in the totality of its parts, and he said 'what is in something *not as a part*'. So 'whole' must be understood as being included *apo koinou*[78] in the
5 phrase 'not as a part'; for not only that which is attributed[79] to something should not be a part, but also the substrate (*hupokeimenon*) should not be a 'part' of what is attributed to it; for the whole is described as being *in* its parts as being predicated of them as subjects.

[77] Spengel's suggestion of *genous* (genus) for *zôou* restores a logical sense here, but it is possible that Dexippus has been distracted by the particular example 'Man-Animal', so the text may stand.

[78] I keep this in the original as a technical term of Greek grammar. There is no compendious English term, so far as I know, for a single word serving two grammatical roles.

[79] The term *epon* is rare in the sense of 'attribute' or 'characteristic', but Plotinus uses *ta eponta* at *Enn.* 2.4.10,27 and 29 of attributes in Matter.

[**24**. That fragrances, although being qualities and 'in a subject' are yet separable.]

SELEUCUS: And how is it not absurd to say[80] that that which is in a subject cannot exist without that which it is in, when the fragrance can be separated from apples and roses? For 10 often the qualities remain after those things which primarily possessed them no longer exist, as when the smell remains after wine has been drunk or garlic eaten?

DEXIPPUS: But he did not say that it was impossible for what is in a subject *to be separated* from what it is in, but *to exist apart*. For the fragrance can be separated from the roses, 15 but it cannot exist apart from them, but either perishes or transfers itself to other subjects.[81] What he says next delineates the position still more accurately; for he did not just say of that alone that it cannot exist apart from that in which it was, but absolutely, apart from everything in which it *is*.[82] The consequence of this is that even if the fragrance formerly in the apple changes its substrate by coming to be in the air, nevertheless it is in some substrate (subject), though 20 in different ones at different times, and never in any way does anything escape the definition of accidents (*sumbebêkota*).

[**25**. A number of problems relating to the peculiar nature of things 'said of a subject', and solution of these.][83]

SELEUCUS: But how are we to counter contentious opponents (*eristikoi*) who, taking attributes that do *not* pertain to the predicate as being 'said of' it, apply to the subject the 'syllogism of negation'?[84] As for instance, 'man' is

[80] This *aporia* appears in Simplicius (49,10-14), not attributed to anyone in particular (*tines*), and is answered rather more elaborately, but to the same effect (49,14-30). Cf. also Porphyry 79,24ff., which is verbally very close, but may have been taken over bodily by Iamblichus.

[81] He means that one can infuse the scent of roses into oils, for example. We are at a disadvantage here once again in face of the range of meaning of the Greek term *hupokeimenon*, 'substrate' as well as 'subject'.

[82] i.e. apart from any particular container.

[83] Dexippus actually only presents one problem in this chapter. The title seems in fact to cover the next four chapters. This problem relates specifically to *Cat.* 1b10-12.

[84] The *sullogismos apophaseos* (or *antiphaseos*, according to mss. AM) seems to run: 'Socrates is a man; man is not (identical with) Socrates; hence Socrates is not Socrates.' Exactly who put forward such an argument originally is not clear. Perhaps Megarians or Pyrrhonists? It is mentioned neither by Porphyry nor by Simplicius. It

25 predicated of Socrates as subject, but of man also there is predicated 'not being Socrates'; and so 'not being Socrates' would be predicated of Socrates.

DEXIPPUS: Once again, in face of these also, we take into account, not as the Stoics propose, the exclusion of negative premises, but as Aristotle teaches, the assumption of 26,1 essential predicates (only), a distinction which the others do not observe, and so fall into logical error, applying non-characteristics to a substance (*ousia*) as if they were being said of it as a subject.[85]

[**26**. The genus is not 'said' of the species as one thing of another.]

SELEUCUS: If someone were to claim that the genus is not 'said of ' the species at all as one thing is said of another (*kath' heterou*), since it is predicated of its subject according the 5 same account (*logos*)[86] and in so far as it shares with it in a common essence (for only in thought (*epinoiâi*) do we separate genera from species – they do not subsist independently, but have their being in the species), how are to we answer?

DEXIPPUS: Surely that the apparent sameness of these actually conceals a difference in relation (*skhesis*), as the non-related differs from the related;[87] the genus is not yet

may nevertheless be an *aporia* of Lucius and Nicostratus.

[85] Busse's text is confused here. Presumably we are to read *hôs kath' hupokeimenou legomena kat' ousian proslambanontes*. Dexippus' solution to the *aporia* appears to be as follows: the *aporia* shows that the principle of *Cat.* 1b10-12 must be suitably restricted so as to block such syllogisms as that produced here. Presumably the Stoics wanted merely to exclude negative minor premisses, but Dexippus claims it is more Aristotelian instead to refuse to allow any non-essential predication as the minor (I am indebted to Steven Strange for this elucidation).

[86] Reading *kata ton auton logon* for *kata ton autou logon* of mss. (Busse reads *hautou* for *autou*).

[87] That is, related to a definite object; *akatataktos* and *katatetagmenos* are used in this sense in Simpl. 27,15-24, and 53,7-9. This latter passage shows that Dexippus is sticking more closely to Porphyry (in his big commentary and not, strangely enough, in his short commentary; cf. 80,29-81,2) than to Iamblichus. Simplicius reports Porphyry as replying to the *aporia* (which actually adduced Aristotle's own example of 'man' and 'animal') as follows: 'There are two senses (*epinoiai*) of "animal", the related and the non-related; the non-related is predicated of the related, and in this way it is "other".' Iamblichus (ibid. 53,9-18) is more elaborate, and not reflected in Dexippus. It may be that Dexippus felt that he was not adding anything substantial to Porphyry's explanation, which would indeed be a fair assessment. Iamblichus says: 'It is not the genera that are predicated of the subjects, but other things by means of them; for when we say "Socrates is a man", we are not saying that he is the generic

related, as for instance would be the case if one were to consider man as such, whereas the species is already related to its particular differentiae, as when we think of the citizen as a member of a particular tribe (*phuletês*). 10

[**27**. Why is 'man' said of Socrates, but the 'species' (*eidos*) no longer so?]

SELEUCUS: How comes it that 'man' is said of Socrates, and 'species' of man, but 'species' is no longer predicable of Socrates; and 'animal' of man, and 'genus' of animal, but 'genus' is no longer predicable of man?[88] 15

DEXIPPUS: Because the higher terms are predicated of the subjects by name, but not by definition (*logos*); for neither, if one is replying to the question 'What is man?', will one say that it is the species, nor does the definition of the species fit the term 'man'. And the same goes for the genus. Furthermore, the species and the genus are predicable of different things and are contradistinguished, the latter by reason of being not applicable to species, the former by not being applicable to individuals, whereas what are said of a subject are constitutive of its essence. And further, some things are said of *all* of a subject; e.g. every man could be reasonably said to be an animal, because there is nothing which falls under the description 'man' of which one cannot say that it is an animal; but to say that every man is a species would be false, and it is unsound to say that every animal is a genus; for the individual man is a man, but not a species, and the particular animal is an animal, but not a genus. So these are not said of all the given class, but will only be said of them in the way that commonalities (*koinotêtes*) are, and because 20 25

Man, but that he participates in the generic, even as saying that a vine is "white" is the same as saying that it bears white grapes, where the vine is so-called by reference (*anaphora*) to its fruit.'

[88] This *aporia* is dealt with by Porphyry, 80,32-81,22, and by Simplicius 52,9-53,4. It is generated by the carelessness of Aristotle's own formulation in *Cat.* 1b10-15, since he actually does not distinguish between the relation of an individual to its species and that of a species to its genus, thus leaving his exegetes with a problem which taxed their ingenuity, namely, to specify what attributes of the predicate in essential predication will also apply to the subject, as the unrestricted application of Aristotle's stated principle does not do this, and is thus false. Dexippus' solution is similar to that found in Porphyry and Simplicius, but it is not verbally close to either.

the one is constitutive of the essence, while the other is
30 indicative of the imposition of the name denoting a common relation (for this commonality that is predicated is not man or Socrates).[89] This latter, then, seems rather to be indicative of the name, indicating in what context (*skhesis*) 'man' or 'animal' is being named, and some names are secondary to other, primary ones, but do not depend on the principal meaning of the expressions (*lexeis*), nor do they signify the
27,1 actual nature of the thing named, but indicate the characteristic (*kharaktêr*) according to which they are named.[90]

[28. In what sense does he say that some differentiae of the predicate are predicated also of the subject, and in what sense all, and what ones can do either?]

SELEUCUS: But there seems to be a degree of contradiction in the arrangement he now makes. For when he says (1b21) that 'there is nothing to prevent genera subordinate to one
5 another from having the same differentiae', he is stating what is the case with some, but not absolutely with all. However, when he adds the further definition that 'the higher are predicated of the genera below them', and 'all differentiae of the predicated genus will be differentiae of the subject also', he seems to be making a statement about all of them at once, without exception.[91]

[89] I place a full stop here, and a comma earlier, after *rhêthêsontai* (l. 29). Porphyry, in a parallel passage (81,18-20) says: 'Man differs from "Socrates" in that the former is used as a universal, whereas the latter is not used as a universal, but as a particular.' Dexippus' point here is rather different, that a word like 'species' describes a sort of *koinotês* quite different even from 'man', which can be applied to an individual subject, like Socrates.

[90] Dexippus is here, it would seem, relying on a distinction he has made earlier (15,16ff.) between words of 'primary' and 'secondary' signification (cf. n. 38 above).

[91] This is indeed a notorious problem. Ackrill comments (*Comm.*, p. 77): 'The last sentence (16,20-4) probably requires emendation. As it stands it is a howler, unless we take "differentiae of the predicated genus" to refer to differentiae that divide it into sub-genera (*differentiae divisivae*) and "differentiae of the subject genus" to refer to differentiae that serve to define it (*differentiae constitutivae*). But there is nothing in the context to justify such an interpretation. Only *differentiae divisivae* are in question.' Ackrill goes on to point out that a correct statement would be obtained if the words 'predicated' and 'subject' were transposed. This is actually proposed by the Peripatetic commentator Boethus, ap. Simpl. 58,27ff. (cf. also Porph. 84,34ff.), but it is hardly plausible that a later editor or scribe would perform a transposition that creates such a 'howler'. Better is Simplicius' solution (described, without

DEXIPPUS: It is clear that what refers to some and what refers to all are not being defined in the same way, but differently; for since there are some differentiae that produce the species (*eidopoioi*),[92] by virtue of which the genus is distinguished into species and qualified, e.g. animal being qualified by 'ensouled' and 'endowed with sense-perception', which are qualities which constitute its essence, and there are others that divide it (*diairetikai*), by virtue of which the genus is divided into the species subordinate to it, e.g. 'animal' into 'rational' and 'irrational', it is obvious that the differentiae which apply essentially to 'animal' *a fortiori* apply also to 'bird', while those which divide it into the species contained by it will in some cases apply to 'bird', when they are relevant (*koinônein*) to it, and sometimes will not, when they are not, as for example 'rational'. 10 15

It is true, then, that both some and all of these differentiae of the predicate are also of the subject, and which these are we have now sorted out. It is not therefore necessary to emend the text by changing round the elements of the statement as follows: 'so that all the differentiae of the *subject* genus will be differentiae of the *predicated* genus also.'[93] This emendation is misguided; for it is not the case that any given differentiae of 'man' could be applied to 'animal', I mean, 'two-footed', and 'capable of discourse' and so on, but it is plain that the expressions 'there is nothing to prevent' and 'not all being predicated' refer to the divisive differentiae, while 'all being predicated' refers to the constitutive ones. It seems to me, at any rate, that he himself indicates this when he distinguishes 'higher' (*epanô*) from 'lower', not making use of such a distinction only in relation to genera, but also in relation to 20 25

acknowledgement, by Ackrill above) that we must distinguish between *diaphorai diairetikai* and *diaphorai sustatikai*, 'differentiae of the subject genus' being the latter only. In fact, the statement at 1b20ff. follows directly from the principle of 1b10-12, which in the form in which Aristotle states it, is, as we have seen, false.

[92] This term is an indication that Aristotle's discussion of differentiae in *Topics* 6.6 (cf. 143b7) is an influence here. Cf. Porph. 85,11ff., showing that this solution goes back to Porphyry, at least.

[93] The proposal of Boethus, see n. 91 above (alluded to also by Porph. 84,34ff.). Boethus did also suggest, it must be said, what is probably the real solution, that Aristotle is intending his final sentence to be dependent on the *ouden kôluei* of 1b21, with the result that all he means is that *in those cases* where differentiae of *hup' allêla* genera are the same, the differentiae of the predicates will be the same as those of the subjects, a perfectly harmless statement.

30 differentiae. So we will take it that the 'higher' differentiae here referred to are those which are constitutive of the relevant essences, such as one can predicate of the subjects also, and the 'lower', such as are confined to the genera prior to them, which sometimes may not be thus predicated. One might also make the distinction by terming the one set
28,1 'generic' differentiae, and the others 'specific', the specific being such as divide the genus into parts, in virtue of which the species subordinate to the species are 'specified', and the generic, in turn, those differentiae which are constitutive of the genera, in virtue of which the genera are 'specified'.[94] There are, however, some of the 'divisive' differentiae which can pertain to the subjects,[95] as has been shown in the case of
5 species which are subordinate to each other.

[**29**. That some distinct genera employ the same differentiae.]

SELEUCUS: What would you say about the fact that distinct (*hetera*) genera employ the same differentiae? For instance, 'equal-sided' and 'unequal-sided' are each equally called figures, and yet 'figure' is a name common to both of them, and virtues and vices, though differing in genus, have the same
10 differentiae; for 'rational' and 'irrational' are predicated of both of them.

DEXIPPUS: People who raise this difficulty have fallen into error by reason of their ignorance of what it means to be a distinct genus. For even if these are not subordinate to each other, they are still not 'distinct', seeing that they are ranked under one common genus, e.g. 'straight-line figure', which in

[94] This terminology is that of Iamblichus, as we learn from Simpl. 59,33-4. From the way in which it is introduced here, and from the similarity between ch. 28 of Dexippus and 84,26-86,4 of Porphyry, we might conclude that Dexippus has used primarily Porph. and only secondarily Iambl., but the alternative is that Iambl. himself transcribed Porph., with whom he has here no quarrel.

[95] The mss. here read *tois autois*, 'the same', which makes very little sense. Fortunately, we have a closely parallel passage of Porphyry (86,3-4), to which Busse draws attention, which makes the sense clear: 'There are, however, some of the divisive differentiae that lie below animal which can come also to belong to the subject in dividing it, as we have shown.' In this case, unlike 26,31 above, I venture to alter the text, reading *tois hupokeimenois*, as Busse suggests (Diels's *tas autas* would be another way out, palaeographically more acceptable, meaning, 'which can be *the same* (in both cases)').

turn is subordinate to 'figure';[96] and virtue and vice similarly will be ranked under the genus 'habit' (*hexis*), and such genera will thus not be distinct from one another. 15

[**30**. Problems relative to heterogeneity.]

SELEUCUS: Similar to these also are problems like the following: they say that the footed and the winged and animal and bird have a number of differentiae the same, so that it is not only in the case of homogeneous entities and genera subordinate to one another that there are sometimes differentiae in common, but also in the case of separate genera (*diestêkota*); and so what was said just now will be false.

DEXIPPUS: To this we may make the same reply. We will say, 'you misinterpret the sense of "distinct" (*hetera*) genera; for by "distinct" he does not mean ones that vary slightly from one another, but *completely* separate ones'. It is the same when they say that 'footed animal' and 'winged animal' have the same differentiae; for both of them may be grass-eating or meat-eating or seed-eating,[97] although not being subordinate to each other, but contradistinguished. So it is easy to make the same response to this difficulty also: 'footed' and 'winged' both share the genus 'animal' and therefore are to be taken as completely distinct genera. 20 25

[**31**. What is a species, and what is a differentia, and how are they to be distinguished from one another?]

SELEUCUS: Well now, would you give me a brief account of species and differentia?[98]

DEXIPPUS: Species has an element of its account in common (with its genus),[99] but is separated off from its proper genus by its specific differentiae, and has as a distinctive element in 30

[96] There is some small corruption here. I read <*ho*> *kai hupo to skhêma*, following a suggestion by Diels, who would, however, substitute *ho* for *kai*.

[97] cf. Porph. 84,10-20, and Simpl. 59,16-19, though these examples are there being used in a rather different context.

[98] Simplicius discusses this topic much more fully at 54,24-56,15, incorporating the definition first of Porphyry (in his *Isagoge*) and then of Iamblichus, but none of it, except one sentence noted below, is verbally very close to what we have here.

[99] If that is the meaning of *koinou ekhetai logou*.

29,1 its definition the quality of having an account (*logos*) in common with individuals, even as the genus has an account in common with its species. The differentia is that which has the role of separating off from things that fall under the same genus those things which are characteristic of the particular species, and which indicates how each of the species is different; for in respect of genus they exhibit no difference. However, not everything that makes a distinction is *ipso facto* 5 a differentia; it must make an essential distinction; for example, it is not the case that if sailing differentiates us from the other animals, 'sailing' (*to pleustikon*) will therefore be a differentia of man;[100] for this quality does not make a distinction which relates to man's essence (*ousia*), but it is simply an accidental characteristic. So the differentia requires to be part of the essence; for it is in relation to their essence that it defines the distinctiveness (*idiotês*) of those things that it separates off. Differentia and species, then, 10 differ as whole from part; for even as is the relation of the word (*onoma*) 'rational' to its definition, so the thing (*pragma*) corresponding to the differentia is comprehended in the species.

[**32.** How does one discover the genus and the species and the differentia?]

SELEUCUS: Since in all areas of study it is of the utmost necessity to know the modes of discovery of the subjects one has set out to investigate, here too, in the present case, let us try to set out their mode of discovery in some scientific way.[101]

15 DEXIPPUS: That the genus and the species are discovered on the basis of an essential commonality (*koinotês*) is obvious enough. And this also is plain from our previous definitions, that the species extends by virtue of its common characteristics throughout the individuals (under it), while the genus

[100] This example occurs in Porph. *Isag.* 12,4-10, and, taken either from there or from Porphyry's large commentary, in Simpl. 55,10-12. Dexippus and Simplicius share the adjective *pleustikos*, which is not found in the *Isagoge*, but may well have been used by Porphyry elsewhere. The intermediacy of Iamblichus should also not be excluded. The passage 29,5-10 as a whole is paralleled closely in Simplicius (55,9-13), though not quite verbatim.

[101] There is nothing corresponding to this enquiry in either Porphyry or Simplicius. It may be a contribution of Dexippus himself.

extends over the relevant species, and is on a still more universal level. The differentia we discover from the characteristic distinctions between the species. It holds the intermediate position between substance and quality, either 20 through being compounded from both of these, or through substance taking on quality, or the other way about, quality taking on substance, or by the differentia being somewhere in between the qualities and the substance, so as to transcend both these extremes.[102] So then, the account (*logos*) of differentia should never be dissociated from the essence (*ousia*) of the species (otherwise it becomes a mere accident), but even as the genus is discovered from the conjunction of the 25 species and the substance, and the species, whenever individually countable (*arithmêta*) entities come to have an essence (*einai*) in common, so also differentiae are discerned from the perceived distinctness (*heterotês*) in the characteristics of a given substance; for thus would their discovery be easily achieved.

[**33**. What does he mean by 'the differentiae are themselves different in species (*eidei*)'?]

SELEUCUS: Would you explain the passage of text (*edaphion*) where he says 'of genera which are different[103] the differentiae are also different in species'[104] (1b16-17). 30

DEXIPPUS: He is using 'in species' in the sense of 'in account' (*en logôi*), as it is also his custom in the *Topics*[105] to interchange these two. His purpose in this is to attain accuracy on a larger matter; for in order that we may not, through paying too much attention to terminology, attribute 30,1 the same differentiae to a piece of furniture and a living thing,

[102] A piece of inspired waffle by Dexippus on the notorious problem of the categorial status of differentiae.

[103] Dexippus seems to have read here *heterôn genôn* for *heterogenôn* of Aristotle's mss. (mss. of Dexippus, A and M, read *heterogenôn*, but cf. 30,4 below), as does Simplicius (54,22, etc.), but not, interestingly, Porphyry in his short comm. (81,26 (lemma), 83,7; 84,4).

[104] *En eidei* here would be more idiomatically rendered 'in kind', as indeed Ackrill does in his translation, but it seemed best to keep the word 'species', since the *aporia* arises from the ambiguity in Aristotle's use of the word here.

[105] Busse's ref. to the *Topics* (1.7, 103a8ff.) seems hardly apposite, since there we have a distinction *arithmôi* – *eidei* – *genei*, where *eidos* precisely does mean 'species'. However, I have nothing better to suggest.

e.g. 'footed' and 'footless',[106] for this reason he has added 'in account'; for these appear to have the same differentiae only by reason of the name. We could put the case rather thus: as those differentiae are 'different in species' which are taken as belonging to species contained in each of a number of different
5 genera, man being taken as in the genus 'animal', and music in the genus 'knowledge', the expression 'different in species' can signify the specific (*eidopoioi*) differentiae, so that one may not take accidental ones, such as 'sailing' of man; for it is not just any sort of differentiae, but only those that make a specific and essential distinction, that are properly 'different in species'.

[34. What is the meaning of 'atomic' (*atomos*) and 'numerically one'?]

10 SELEUCUS: It remains for us to define accurately what is meant by 'individual' and 'numerically one'.[107]

DEXIPPUS: 'Atomic' is used, not in the sense of 'indivisible' (*adiairetos*) or 'partless' but in the sense of not admitting of division as of genus into species, or of species into things distinguished numerically as one. Each one among sensible objects is said to be one not by reason of a single potency
15 (*dunamis*), but because all converge in a single substance. But since 'one' is an ambiguous term (*pollakhôs legetai*), signifying one in genus or in species or in a number of other ways, he now defines one by taking it as one *numerically*, not just in the sense of the number by which we count, as some have maliciously suggested, but also the number counted. For thus we may talk of a group of five horses as being a single number.

[35. How is it possible for what is one and individual to be distinguished numerically from what is one and individual?]

20 SELEUCUS: But if a species is that which is predicated of a multiplicity of numerically different entities in the mode of

[106] These examples are no doubt provoked by the *dipous* of 1b20.

[107] This question takes us back to ch. 2, 1b6-7, by association of ideas from *tôi eidei* of 1b17 (cf. Simpl. 51,7-15, though there is no very close connexion), but in fact it follows logically, in Dexippus' scheme, upon chs. 31 and 32. After defining what is meant by genus, species and differentia, we need now to define 'individual'.

essence (*ti esti*),[108] in what way does one thing that is individual and one differ from another thing that is individual and one? For both the one and the other are numerically one.

DEXIPPUS: Those who seek to solve this problem by reference to the concept of 'individually qualified object' (*idiôs poion*),[109] i.e. by claiming that one person, for instance, is definable as 'hooked-nosed', 'fair-haired', or by some other conjunction of qualities, another as 'snub-nosed' or 'bald' or 'grey-eyed', and 25 another again by other qualities, do not seem to me to solve it properly. For it is not the conjunction of qualities which makes them differ numerically, but, if anything, quality as such. We should rather reply to the problem as follows, that things that are numerically distinct do not differ from each other in nature and essence, but their distinctness resides in their countability. They are different, then, in being countable; for it 30 is in the process of each being counted one by one that number arises. So, it seems to me that to say that things are numerically different implies 'separate (*diestêkota*) so as to be countable', so that what is being said about the species is somewhat as follows, that the species is what is predicated of a multiplicity of entities separate so as to be countable, so that we may take 'different' in the sense of 'separate'.

[**36**. That the division into Categories is not excessive.][110]

SELEUCUS: I have heard some people criticising this 35 division (into categories), on the grounds that he did not do 31,1 well to set off 'acting' from 'being acted upon'; for they ought both to have been brought under the single genus of Motion.

DEXIPPUS: The solution to this is as follows: neither is that which acts, in so far as it acts, acted upon, nor does that which

[108] This definition occurs in Porphyry's *Isagoge*, 4,11-12.

[109] A Stoic concept; cf. *SVF* 2.395-8, and J.M. Rist, *Stoic Philosophy*, 160-4. Stoic common and individual qualities have actually been mentioned before, at 23,28-9.

[110] We now turn to a series of problems arising from ch. 4. Cf. Porph. 86,22-4 and Simpl. 60,4-10. Among those criticising the basis of Aristotle's division Porphyry mentions the Stoic Athenodorus' work *Against Aristotle's Categories*, and Cornutus, in his *Handbook of Rhetoric* (*Rhêtorikai Tekhnai*) and his *Reply to Athenodorus*. Simpl. (62,24ff.) gives this title as *Reply to Athenodorus and Aristotle*, and goes on to add Lucius and Nicostratus. Athenodorus, at least, criticised the multiplicity of categories, presumably arguing for the Stoic system. To all these may be added Plotinus, who criticises the categorisation of 'acting' and 'being acted upon' in *Enn.* 6.1.15ff. (cf. Simpl. 306,13ff.).

is acted upon, in so far as it is acted upon, act, nor in the case of composite entities do the two come together by reason of the confluence of the two principles into the same place. Being moved, after all, is in the sphere of being acted upon, while that which acts moves while itself being unmoved. Some active entities among sensible things move incidentally by reason of the fact that they contain both principles within themselves, and one element in them is motive, while the other is moved. So he distinguished 'being acted upon' from the active agent in accordance with his views on causal principles.

SELEUCUS: Again, another accusation of excessiveness is laid against him on other grounds. They make a division of things into Absolute (*kath' hauta*) and Relative (*pros ti*),[111] and these they regard as taking in all the categories. Others make a division into Substance and Accident, and consider that that is a sufficient distinction.

DEXIPPUS: Well, they are wrong. Those who divide the categories into Subject (*hupokeimenon*)[112] and Accident touch to some degree upon Aristotle's primary division into the least number of genera, and to this extent they are not in error, but in so far as they leave aside the universal and the particular, in this respect they show a poor and defective grasp of Aristotelian doctrine. Those in turn who make a division into Substance and Relative (*pros ti*) – since accidents are *of* something else – when in this way too they make all the categories into two classes only, they bring on themselves the accusation of deficiency of terminology. For not all things are *of* something else by reason of being accidents, but one only, that which is called relative (*pros ti*). All accidents seem to be relatives because of our habit of leaving out words in those phrases where the meaning can easily be derived from the context because of its familiarity. Thus, 'whiteness' seems to be *of* something, although it is not *in itself* of anything. By this reckoning even things primarily absolute are 'relative'. For Socrates is said to be *of* Sophroniscus, though he is not so in so

[111] These Simpl. identifies (63,22) as 'Xenocrates and Andronicus and their followers', i.e. Andronicus, appealing to the Old Academic system as propounded by Xenocrates.

[112] It is curious here that Dexippus in his reply substitutes for *ousia* *hupokeimenon*, the Stoic term for Substance.

far as he is Socrates, but we leave out, as being a familiar convention, 'son', since *qua* Socrates he is 'of' nobody. And if we relate the accident to a given subject, it is a relative (as for instance the farm, as a possession, is a relative), but in and of 30 itself the accident is a distinct nature. And indeed, matter and the substrate, in so far as it qualifies for these epithets, is a relative (for it is substrate to something), while as a principle 32,1 it is distinct (*kata diastasin*) and absolute.[113] We must, then, become examiners, not of linguistic idiom, but of the true nature of things, and not pick on the inadequacy of names (*onomata*), but accommodate ourselves to the differing peculiarities of things (*pragmata*). In that case the genera will not be found to be less than ten. But if one makes a division 5 into subject and what relates to the subject, and on this basis indicts the division into ten as excessive, such a person is actually making an (improper) division between the single, definite subject and 'what relates to it', which is an indefinite term .

[**37**. That the division into categories is not defective.]

SELEUCUS: So much is sufficient for the present in reply to those who maintain that the distinction of the categories is excessive, but I would like you to meet the objections of those 10 who declare it to be defective. For why, they say, did he distinguish 'doing' from 'being affected', whereas he did not distinguish between 'having' and 'being had'?[114]

DEXIPPUS: We would say by way of justification that he did not simply omit the latter distinction, but (he left it out) because it fell under the just previously identified category of 'being in a position' (*keisthai*). For someone *has* his shield, but 15 the shield is 'had' by being in a certain position; position (*thesis*) is therefore nothing else than the arrangement (*taxis*) of things that are had.

[113] Everything up to here Simplicius, in the conclusion of his parallel passage (64,3-4), identifies as the views of Porphyry and Iamblichus. What follows in Dexippus, down to 32,8, is presented, remarkably, by Simplicius (64,4-12, not verbatim, but in substance) as his own view (*mêpote de* …). It looks in this case as if he may have borrowed something from Dexippus himself!

[114] This is declared by Simplicius (64,13) to be an *aporia* of 'Nicostratus and his followers', and he answers it in terms closely parallel to this.

SELEUCUS: They also raise the question as to why he left out conjunctions.[115]

DEXIPPUS: Because, we say, the employment of them is not a primary but a secondary use of language, nor are they complete, but incomplete, nor really parts of discourse (*lektikê*), but act as symbols (*sumbolikê*); nor do they signify
20 primarily, but rather in a subsidiary way, even as we are accustomed to use marks of punctuation (*diplai*),[116] which in combination with the text contribute to the signifying of breaks in the thought, but on their own they mean nothing. So also, then, conjunctions signify in a subsidiary way, in combination with the other parts of speech, but they themselves are not significant on their own, but are like glue.
25 It is for this reason that we do not class them as elements of speech, but, if anything, as parts of speech. Even if these do signify, they signify only in combination, like, for example, the syllable 'ba', and we say that the present subject of discussion is words without combination which are significant by themselves, and for the primary uses of language, not the secondary ones.

30 SELEUCUS: Again, some people raise the question as to where the articles (*arthra*) are to be ranked.[117]

33,1 DEXIPPUS: It is the same story here as well. For they have the force of conjunctions in the area of (identifying) male and female, along with helping to indicate the subject indefinitely (for which reason they are termed by some 'indefinites');[118] for the word 'a' (*ti*) indicates a subject, but when connected with a

[115] This Simplicius presents (64,18) as an *aporia* of Lucius (not Nicostratus?). His reply here also follows closely the text of Dexippus – indeed it required supplementation from it, as it contains a lacuna at 64,20. Despite the closeness, Simpl. is probably transcribing from Porphyry or Iamblichus.

[116] *Diplai* are really marginal marks, used by grammarians and scribes to indicate such things as a variant reading, a rejected verse, or a change of speaker, but this translation preserves the sense well enough.

[117] This *aporia* too is recorded by Simpl. (without attribution, but presumably by Lucius and Nicostratus), though much more summarily (64,29-65,2).

[118] Who this can have been is not clear. Pronouns are called *aorista* by Apollonius Dyscolus (*Pron.* 7,1, etc.), but not *articles* (though what Dexippus means by *arthron* seems to comprehend what we would call pronouns, such as *tis* or *hode*). Such a term would apply properly, one would think, only to the indefinite article (if we take *tis* as performing this role), but then *tis* does not distinguish between male and female. This, however, does not seem to bother Dexippus. (It can be argued, admittedly, that the definite article is *relatively* indefinite. When one says 'the man', one does not yet specify which man. I am indebted for this observation to Steven Strange.)

male entity it indicates something masculine, and with a female, something feminine. The word 'this' (*tode*) is derived 5 from 'a' (*ti*) and either of the articles helps, in combination, to indicate the male or the female, but on its own it is meaningless (*asêmon*).

SELEUCUS: But, they say, where will you rank negatives and negative expressions, indefinites and derivative forms in each category?[119]

DEXIPPUS: On this subject Aristotle himself has instructed 10 us better in his *Notebooks*;[120] for by adding to the predicates (*katêgoriai*) the specification 'with their cases (*ptôseis*)',[121] he has included with them negatives and negative expressions and indefinites, using the term 'cases' to cover derivative forms in general. Thus 'being unshod' will be of the same category as 'being shod', and 'being unarmed'[122] of the same 15 category as 'being armed'.[123] So negations, then, will be of the same category as their corresponding positive forms, and the indefinites of the same category as the corresponding definites; for in general negations and negative expressions derive their existence from the fact that what they negate do not exist,[124] so that it is reasonable that they should be referred to their correspondent existents and positive expressions. So a corpse will be of the same category as a living thing, and blindness of the same as the possession (of 20 sight), and so on in all other cases.

As regards the number 'one',[125] that it does not fall outside

[119] This *aporia* is dealt with by Simplicius, at 65,2-13. Of the various terms mentioned, *apophasis* refers to negative particles (*ou* or *mê*), *sterêsis* to negative words (e.g. prefixed with *a*-), *aoriston*, this time, to negative expressions such as 'not-man' (cf. Ar. *Int.* 16a32), and *enklisis* to cases of nouns and tenses of verbs.

[120] *Hupomnêmata*. Simpl. mentions these also, specifying them further as *Methodika*, *Diaereseis*, and a treatise called *On Speech (ta peri tên lexin)*, of which the genuineness is doubted (65,4-7). This passage is given by Rose (3rd ed.) as Fr. 116.

[121] Dexippus is somewhat garbled here. I translate with help from Simplicius (65,8). This passage is a good indication that Dexippus (if his text is not corrupt) is summarising carelessly a text which Simplicius is quoting more fully.

[122] Dexippus' text gives *anapnein*, 'breathing', which is quite inapposite. I assume, with Busse, *anoplein*, being 'unarmed', a word otherwise unattested (possibly a nonce-word formed on the analogy of *anhupodetein*) but represented by Felicianus' translation *exarmatum esse* (though Felicianus may be simply emending his text).

[123] i.e. 'having' (*ekhein*).

[124] Simplicius says 'from the fact that they negate existent things' (65,12), which makes better sense.

[125] There really should have been a question by Seleucus here. Simplicius in the corresponding place (65,13-17) gives an *aporia*, as follows: 'But, they say, the One and

the categories one might demonstrate in many ways. For in so far as it is the first principle of number and as measure it will be ranked among relatives, and if one were to take number to be incorporeal and absolute (*kath' heauton*), the monad would be a substance, whereas if one postulates two sorts of number, 25 then the monad will be found in two categories, on the level of composite numerables (*arithmêta*), ranked among relatives, while on the level of intelligibles, in the category of Substance. If one were to reckon it as a part of Quantity, as does Alexander,[126] it will also at the same time be in the category of Quantity. However, this whole line of argument is leading us astray; for in all cases the limit (*peras*) is other than what is 30 defined by the limit, and so the monad inasmuch as it is first principle and limit of quantity, would not itself be a quantity. 34,1 But perhaps it is better to leave 'one' also as a homonym – as in many other cases, as we were saying just now,[127] there would not be a primary division into the categories, nor a definite placing of them (sc. homonyms) into any one class.[128]

[**38**. That he does not introduce some classes of categories instead of others.]

SELEUCUS: That is a reasonable reply to those who say that this distinction of his is defective, but how would you deal 5 with those who say that Aristotle introduces certain classes of category instead of others (which he should have introduced?).[129] Some people say, for instance, that Motion should

the monad and the point, how do they not fall outside the categories? For it is not a quantity, as it might seem to some' and so on. From the text, it looks as if Dexippus has for some reason omitted the *aporia*, rather than that there is a lacuna in the text.

[126] Dexippus is oversimplifying somewhat here. The parallel text in Simpl. (65,17-24) tells us that Alexander held that 'as first principle of numbers and as measure, "one" should be ranked in the category of Relative'. It is Boethus who makes the twofold distinction of number (65,19-21), and suggests that the lower level might be better considered as a quantity. But then (65,24ff.) Alexander is reported as holding that the monad is 'a part of Quantity', because 'a number is a multiplicity of conjoined monads'.

[127] Reading *elegomen* for *legomen*, and taking this as a reference back to 13,7ff.

[128] This is incoherent as it stands in Dexippus' text, but receives clarification from Simpl. (who refers to it as the view of 'the more distinguished among the commentators', sc. Porphyry and Iamblichus, 66,12-15): 'It is better to leave "One" as homonymous over all the ten categories, even as is the case with "being"; for in the case of homonyms there is no definite division nor referral to any one class.'

[129] This whole section, to 34,24, is closely paralleled in Simpl. 66,16-67,8, though Simpl. is somewhat fuller. For the objection, cf. Plot. *Enn*. 6.11.16-17.

be a category instead of Doing and Being Affected; for neither should it be subsumed into any other class, nor any higher class predicated of it essentially from above, nor can it be ranked either with 'doing' (for many motions fall into the category of 'being affected'), or yet with 'being affected', because many motions are activities. 10

DEXIPPUS: In response to this, we would start by specifying that potentiality and actuality are to be viewed homonymously throughout the ten categories. For this reason we will place neither potentiality or actuality among the categories; for of things that are homonymous and which change their essence entirely there can be no common genus. Wherefore, since Motion proceeds from potentiality to entelechy 15 differently according as it is in the area of Quality or Quantity or any of the other categories, it is not possible to conceive of it as a single category, by reason of its homonymity. And if one were to bring up the complementary case of Rest, we would make the same reply, making also the further point that Rest belongs properly not to things of generation but rather to the intelligible realm. But if one is going to rank most of the 20 categories as belonging to the class of 'Disposition' (*pôs ekhon*), as the Stoics do, it must be demonstrated to them that they are leaving out the great majority of existent things, things in place and in time,[130] quantities in terms both of number and magnitude, 'being shod', and other such classes; for none of such things is classifiable under 'disposition'.

[**39**. That being (*to on*) is not a homonym, but a synonym.]

SELEUCUS: But someone might say that Being is a 25 synonym; for after all, they say, the account (*logos*) and the name (sc. of 'being') are fitted to (all) beings; for each of beings is either active or passive or both,[131] so that if Being partakes of the same name and account, each (use of the term) would be synonymous.

[130] This criticism, at least, seems inapposite, since Place and Time for the Stoics were incorporeal, and their system of 'categories' (the precise role of which is, admittedly, obscure) was concerned with the classification of corporeal entities (*sômata*). Cf. Plotinus' criticism at *Enn.* 6.1.30.

[131] This is Plato's definition of true Being from *Sophist* 247D-E, though it was also that of the Stoics; cf. *SVF* 1.85.

DEXIPPUS: It is easy to reply to this that it is not the
30 definition (*horos*) of 'being' to be 'either active or passive or both'; for such a rendering of the account comprehends the indicated objects as beings, but does not indicate *what* they are. For even as the account of 'dog' is not 'land' or 'sea' or 'heavenly animal', but that which shows its essence (*ousia*), so neither would the above-mentioned account be a definition of Being, especially as this account is not even true of all the
35 species of being. For the definition of the genus should fit the
35,1 species also. So, even as the account 'animal' fits man as well, so this account should have fitted each of the species of Being, like 'dog' or 'man' or 'horse', but no one, when giving a definition of 'horse', says that it is 'either active or passive or both'.

[**40**. That 'is' is either something we predicate of each category, and so we will proceed to infinity, or, if it is a form derived from Being, then Being would be a genus.]

5 SELEUCUS: One might raise the question[132] as to how one should understand 'is' throughout each of the categories; for either it has the same meaning as that of which it is predicated, or a different one. But if substance (*ousia*) is the same as 'is', then it is superfluous to predicate 'is' of it; for Substance is implicitly (*dunamei*) Substance, and Quality Quality. And if we are not rambling in predicating 'is' of them,
10 then 'is' will be different from them. But if it is different, it will be different for each category, so that there will not be ten categories only, but another ten also, answering to the senses of 'is' as categorising each of them, and again if of each of these in turn we predicate a different 'is,' we will get an infinite regress. But if there is one common meaning of 'is' as predicated of all of them, and 'is' is derived from (*enkeklitai*) Being (*to on*), then Being would be a genus.
15 DEXIPPUS: It is obvious that this objection is sophistic, but that does not prevent one from providing a solution to the problem. When we predicate 'existence' (*huparxis*) of it, 'existence', and stating that the subject 'exists', does not

132 This *aporia* is not dealt with by Porphyry or by Simplicius. It may be original to Dexippus.

indicate anything distinct from the subject, but in the case of each of the ten bears witness to its subsistence (*hupostasis*) 20 that it is not insubstantial, but subsists. And so the verb 'to be' will be a homonym and will be used in ten different senses, in as many categories as there are beings.[133]

[133] Dexippus' solution is actually quite remarkable and important, as Steven Strange points out to me, because he is clearly talking about something like the specific notion of *existence*, which is notoriously absent or neglected in earlier Greek philosophy.

BOOK 2

39,1 DEXIPPUS: Up to this, my most studious Seleucus, you have both put forward certain problems propounded by the ancients, and raised others yourself. But now I want to move on to the actual genera of predications themselves. On this question, do you want to separate off the problems raised by 5 Plotinus[1] and take them last, or rather to take them in conjunction with the chapters to which they refer?[2]

[1. Is there a division into categories, or simply an enumeration of them?]

SELEUCUS: My preference would certainly be not to break the continuity, but to examine the problems that come up in due order. So then, starting from Substance, I want first to ask about these categories that have been listed; do you understand them as representing a division,[3] or simply an enumeration?

10 DEXIPPUS:[4] We should not think of them as representing a division; for there is no question here of a division of a genus into species. For neither is there any genus common to the ten categories, such as some introduce in the form of 'Being' or 'Something (*ti*)',[5] nor, if there is no genus to be divided, can there be divisions of it. But if it is not possible to postulate either a genus common to the highest genera or differentiae

[1] Sc. in his treatise *On the Classes of Being*, *Enn.* 6.1-3. In connexion with certain of the *aporiai* attributed here to Plotinus, however, see P. Henry, 'Trois apories orales de Plotin' (Bibl.) and 'The oral teaching of Plotinus', *Dionysius* 6, 1982.

[2] That is to say, chs 5-9 of the *Categories*, in which Aristotle takes up each of the categories in turn.

[3] That is, as being the subject of a formal *diairesis*. This question is raised by Plotinus at *Enn.* 6.3.13,11, and by Porphyry 86,11-13.

[4] The whole passage, down to 40,5, is found in a slightly fuller form, but otherwise verbatim, in Simpl. 61,19-62,6, with the tell-tale subscription: 'This is the view of Iamblichus (*tauta men ho Iamblikhos axioi*).'

[5] The Stoic supreme category.

which would distinguish them, it is not possible to call such an enumeration a division. 15

SELEUCUS: Well then, is the division like what happens in an army, as for instance the grouping into regiments (*lokhoi*)? For in that case each regimental commander assumes command of each regiment, and in this case Substance takes control of substances, while the other categories take control of what falls under them?

DEXIPPUS: But if we say this, the class of beings will be a sort 20 of scattered multitude, and like those demes which are brought together out of scattered tribes and phratries.[6] But this destroys the continuity and mutual implications of the system. It is better, therefore, to define the genera according to certain primary characteristics of the categories and to circumscribe 40,1 the account of the essence (*hupostasis*) of each of them on its own, and in this way to distinguish them apart. Or to look at it another way, in so far as accidents are relative to and present in substance, and are conjoined to it in an organised fashion, to this extent we must assimilate them to those things derived from and centred on a single focal entity.[7] 5

SELEUCUS: What, then? Is it to be likened to the division of a whole into parts?

DEXIPPUS: Not at all; for it does not circumscribe the structure of what one might call an aggregate (*sunkrima*) into certain of its parts, but rather it analyses it into more simple and basic elements by employing always more basic conceptual distinctions. Nevertheless, it preserves a structure (*taxis*) of 10 things said with reference to, related to and proceeding from a single source, even as in the realm of existents substance takes the first and most dominant position, and the accidents have their existence in relation to it.

[**2**. A problem raised by Plotinus, to the effect that 'substance' does not signify the same thing in the case of the primary beings and later ones, since there is no

[6] The demes of Attica were composed of tribes and phratries taken deliberately from different parts of the country, to neutralise local rivalries and separatist tendencies.

[7] Using Aristotle's terminology in e.g. *Metaph.* 4.2, 1003a33ff. or *EN* 1.6, 1096b27-8.

common genus in which primary and secondary entities share.]

SELEUCUS: So much, then, for that. But let us turn to consider the first problem raised by Plotinus.[8] He raises the question, if there are two kinds of substance, intelligible and
15 sensible, how can there be one and the same genus common to both of them; for what is common to the essence of both of these? And even if there is something common, it will be something else prior to these, i.e. something neither body nor incorporeal; for otherwise either the incorporeal will be body or body incorporeal.

DEXIPPUS: It is easy to reply to this, that these difficulties
20 are being brought up irrelevantly.[9] For it is not the purpose of this work to discuss (real) beings, nor the genera of primary substance; for it is aimed at young people who can follow only the simpler doctrines. So since his concern in the present instance is with utterances (*lexeis*) whose property it is to be said of substances, Plotinus is not justified in introducing into
25 the study of these (sc. *lexeis*) questions of ontology.

However, I do not think that we should sidestep thus a course of argument well suited to the present occasion, but rather take our start from the philosophy of Plotinus itself and set the present course of argument in the context of his overall position. For he postulates one single genus of substance in the intelligible realm as being a common source of being to the incorporeal forms and thus bestowing being on
30 the whole sensible realm and on the forms in matter. But if this is the case, and the first principle of substance extends the same through all, having a primary, secondary and tertiary level,[10] in accordance with which it provides being to
41,1 some primarily and to others in another mode – so, if everything relates back to it as dependent on it, then the

[8] In *Enn.* 6.1.2. Simplicius deals with this aporia at length at 76,13-79,5, attributing it to Nicostratus before Plotinus. He is not verbally close to Dexippus here. See on this passage the important article of Pierre Hadot, 'L'harmonie des philosophies de Plotin et d'Aristote selon Porphyre dans le Commentaire de Dexippe sur les Catégories', in *Plotino e il Neoplatonismo in Oriente e in Occidente*, Roma (Accademia Nazionale dei Lincei), 1974, 31-47.

[9] *Para tên prothesin*, 'contrary to the stated subject matter', sc. of the *Categories*.

[10] Probably a reference to the 'seconds and thirds' of *Tim.* 41D, or perhaps of the three levels of *Ep.* 2.312E.

sketchy account (*hupographê*) presented here can be seen as indicating also the first principle from which this has fallen into this lowest manifestation.

[3. That this objection can be made on the basis of Platonist assumptions, but that the problem should be solved on Aristotelian terms.]

SELEUCUS: If Aristotle were proceeding on the same assumptions as Plotinus, these statements might have some 5 reasonableness; but as it is, you are presupposing[11] the truth of Plotinus' hypotheses in your defence of Aristotle.

DEXIPPUS: Well now, in dealing with this problem, I will base myself on what is said in the *Metaphysics*.[12] For there are these two substances according to Aristotle, the intelligible and the sensible, and median between those two 10 the physical;[13] the sensible is the composite (*sunthetos*), the physical that characterised by form and matter, and that which is higher than these is the intellectual and incorporeal, which he often calls unmoved but motive, as being the cause of motion which takes the form of life. For this is what Aristotle lays down about these substances in Book 12 of the *Metaphysics*, and here he subsumes the multiplicity of substances under Substance in general. He brought them all 15

[11] Reading *prolabôn*, with AM, for *proslabôn* of Busse's text.

[12] 12.1069a30ff. (though Dexippus does not seem to have read it very carefully).

[13] This passage is profoundly confused. Reference to the corresponding passage of Simplicius (77,4-10) throws some light on the situation. The root of the problem seems to be Iamblichus' determination to use 'Archytas' to elucidate Aristotle's division of types of substance in Book 12 of the *Metaphysics*, in the process plainly misinterpreting 'Archytas'. Aristotle begins (1069a30) by saying that there are *three* types of substance, and then produces *two*, sensible and 'unmoved', but subdivides the first into 'eternal' and 'destructible', which makes three. Iamblichus (whom Simpl. is plainly using here, the use of 'Archytas' being a tell-tale sign) talks of *two* substances, and a third in the middle, rephrasing Aristotle (in the direction, perhaps, of a Xenocratean triadic division of the universe). But he then adduces 'Archytas', who declared that all *ousia* is *phusikê kai aisthêtikê kai kinêtikê*, plainly saying merely that all *ousia* is natural, sensible and in motion. Iamblichus, however, wants to take this as describing a three-way division of substance corresponding to Aristotle's, and from this confusion arises: *phusikê* is taken to denote the (sublunary, perishable) physical world, *aisthêtikê* the realm of eternal, but visible substances (the heavenly bodies), called also *sunthetos* because a compound of the physical and the intelligible, and *kinêtikê* intelligible substance, as being 'motive' of what is below it. It is all very convoluted, and plainly too much for Dexippus, who makes the *phusikê* the mediating substance rather than the *aisthêtikê*. But this is a good passage for seeing the role of Iamblichus as a source for both Dexippus and Simplicius.

together into one system and traced them back to one originating principle (*arkhê*). For it will hardly be that anything else would participate in unity, if substance itself, which has its being in the One, is to be denied that completeness which is attributable to unity. So since intelligible reality is ineffable, he makes use of the name of 'substance' metaphorically and analogically from what is familiar to sense-perception. For all things that share a name
20 share it in one of three ways;[14] either homonymously or synonymously, or one thing having the name primarily, the other metaphorically; homonymously, as the foot of an artefact (*organon*) with respect to other feet of artefacts; synonymously, as a knife is said to be 'sharper' than another knife or a sound than a sound – for a knife is not said to be sharper than a sound, as the sharp is different in two cases;
25 primarily and metaphorically, as in the case of a human foot and the feet of mountains. So, since (intelligibles) are ineffable, he uses the name 'substance' metaphorically (of them), making them knowable through things sensible and perceived by us. For sensible substance will be homonymous with intelligible substance, representing it only by analogy, but it will be synonymous with physical substance, representing it by its very composition.

30 Just as in the *Metaphysics*[15] he refers to a certain incorporeal essence as 'body' – calling it incorporeal, since it proceeds from intellectual substance, but affirming that it is body, since it has already proceeded to the level of sensible substance – so in saying that this substance is synonymous
42,1 with physical substance, but homonymous with intelligible substance, he teaches us about those kinds by means of this kind. Above all (this is possible) because the description of this kind of substance, as that which is neither said of nor in a subject, can be applied to the other two also.

I have answered Plotinus rather dogmatically, excellent Seleucus, but since these solutions are rather profound,
5 please reserve such replies for your philosophically sophis-

[14] This passage (41,19-27) is paralleled in Simpl. (81,7-14), though Simplicius' version is actually somewhat shorter.

[15] Busse could find no reference for this, and nor can I, but cf., perhaps, *DA* 1.2, 405a7, where fire is referred to as 'the most incorporeal of bodily substances'.

ticated opponents, and employ the easier ones against those who understand little of the doctrine, saying that it is Aristotle's usual practice to fit his lectures to his audience. Thus here he brings in none of the higher philosophical considerations which would be irrelevant to the matter at hand: he uses the term 'substance' as a homonym, since his 10 primary intent is to deal with words; nevertheless incidentally he teaches both about words and about what they signify, so that we must not raise irrelevant objections concerning the latter.

[4. A problem raised by Plotinus, as to whether there are the same genera of beings in the intelligible as in the sensible realm.]

SELEUCUS: Well then, we will employ the same arguments also when Plotinus raises the question[16] as to whether there are the same genera of predications at the level of the primal beings and at the level of those subsequent to them, or 15 whether there are more in the intelligible realm and less in the sensible, or vice versa.

DEXIPPUS: This is a similar problem, Seleucus, to that about substance, so that you can base your solution on the declared purpose (*prothesis*) of the *Categories*. However, one might propound another solution also, following the doctrine which Aristotle sets forth in the *Metaphysics*.[17] For there, accepting 20 that there are ten categories in all, he demonstrates that all of them exist in composite entities, which demonstrate their composite nature not least in this respect, that they are all present also in particular entities (for the addition and multiplicity of characteristics is a universal feature of bodies). So then the ten categories are present in sensible objects, so that one might demonstrate, for instance, that they are present in Socrates, but when it comes to intelligibles they are 25 not all present, but only those genera which pertain to the simplest and most perfect entities. And this will not constitute a pretext for declaring the division of genera defective; for it is

[16] At the beginning of his treatise on the *Categories*, *Enn.* 6.1.1,19-22. No trace of this is in Simplicius; perhaps Dexippus is excerpting *Enn.* 6.1 on his own.

[17] The reference is to *Metaph.* 8.4, 1029b23ff.

not in the event that all are not present in that realm that the genera thus distinguished will be incomplete, but rather since natural objects subsist both in the intelligible and in the sensible realm, they appear to be more by reason of the 30 composite nature of (sensible) essence. For this reason the overall account of the categories is not incomplete.

[**5**. A problem raised by Plotinus against Aristotle, as to what common element it is in matter and form and the composite that one discerns that leads one to call them all 'substance'.]

SELEUCUS: Again, Plotinus produces another problem,[18] as follows: Since 'substance' is used in three senses, each of them different, which is he bringing under the present category, that of 'matter' or 'form' or the composite of the two? For all of 43,1 them cannot be implied in the present account; for they do not all have the same meaning, nor are their meanings related hierarchically, so that whichever of them one implies, one will necessarily be rejecting the others.

DEXIPPUS: But in taking one meaning as 'the composite', he 5 has included also those which are comprised in it as parts; for as a general principle, 'when one thing combines with another to form a unity, the one does not reside in the other';[19] so that the three, I mean 'form', 'matter', and 'the composite', have this in common, that they are not 'in a subject'. The three of them are substances, all gaining their meaning from the simple source of the composite.

[**6**. A problem of Plotinus about Substance, what is the primary and simplest concept of it, and how it is possible to explain it?]

10 SELEUCUS: But again, Plotinus raises the problem that it is

[18] *Enn*. 6.1.2, 9-15.

[19] A reference to *Enn*. 6.3.4,32, so that Plotinus in fact supplies by implication a solution to his own aporia. Simplicius makes the same reference at 94,27-8 in a context where Archytas is being referred to, and which may thus be taken to be Iamblichean. (Note that both Dexippus and Simplicius use the compound *sunapartizô* where Plotinus uses simply *apartizô*.)

not easy, when talking of Substance, to say what it is.[20]

DEXIPPUS: Well, if you wish to arrive at the concept (*ennoia*) of it, consider first the process of communication by means of names,[21] because it is confused totalities that strike our notice (*epibolê*) first, while second comes what needs more definition and articulation. These he now adds, in granting to substance as its particularity 'being neither said of a subject, nor in a subject'. That it is not *in* a subject is obvious (for it serves as a subject to everything else and without it everything else has no being), while not being *said of* a subject is most proper to individual substance. For it is because it is indivisible into species that it is said of no subject, while the secondary substances serve as predicates to it as subject. For that which serves as subject to substances would be substance in the primary and proper sense, since if these do not exist, none of the other things can do so either. So then, there being three senses of 'substance', as we have said, he is now referring to the composite, but later he will make mention also of the remaining two senses, when he says that they too are parts of substance.[22] 15 20 25

[7. A problem of Plotinus, as to why Substance is placed before the others.]

SELEUCUS: Plotinus[23] says that it is worth establishing first why Substance is ranked first among the categories.

DEXIPPUS: One might discover two reasons why Substance

[20] Also in *Enn.* 6.1.2, where his final remark is 'But in sum, it is impossible to define Substance: determine its property (*idion*), and you still have not attained its essence (*ti esti*)'. The reply to this question, and the previous one, is provided by Simpl. at 79,6-30, in terms which he declares to be those of Porphyry, and Iamblichus copying Porphyry verbatim (79,30), but the text does not correspond verbally to anything in Dexippus. On the other hand, in Porphyry's short commentary (88,8-22), we have a much closer correspondence. This reveals a complicated web of correspondences which I am not sure how best to unravel.

[21] Taking *didaskalia* here as just 'communication of information'. Cf. 6,19-20.

[22] Presumably at 3a29ff.

[23] I cannot find Plotinus raising this question in 6.1. Busse's ref. to 6.1.25 is surely irrelevant. It is possible, as Henry would claim ('Trois apories orales ...', 236), that this is an *aporia* orally transmitted, via Porphyry, but since the next Plotinian *aporia* (section 9 below) follows immediately on the one previous to this (section 6 above) we may perhaps conclude that this is wrongly attributed by Dexippus to Plotinus, and is in fact of earlier origin (Lucius and Nicostratus?). At the beginning of 6.1.2, at least, Plotinus seems to accept that Substance is correctly placed first (*apo gar tautês pantôs arkteon*).

should be first. One, because it alone serves as substrate (*hupokeitai*) to everything else; for if you take Matter, it
30 serves as substrate to forms, and if you take Form, it serves as substrate to accidents. But we can provide another reason, that it is conceived of as subsisting itself by itself whereas everything else subsists in it, or not without it, but what
44,1 exists of itself from the start is prior to what is with it or not without it. Thus the monad is prior to the dyad, because a monad can exist without a dyad, but a dyad is always accompanied by a monad.

[8. That the account presented of it does not inform us what Substance is.]

SELEUCUS: But this problem is worth raising also,[24] I think,
5 namely that he does not inform us what Substance *is*, but what it is not, and that it is this that makes other things to be known, but not itself, even as if, in giving an account of the concept of 'man', one were to say that it was neither a horse nor a dog, one has in no way further defined 'man' by this negative description than when he was originally undefined; since not even if one states some proper characteristic (*idion*)
10 of man, e.g. 'capable of laughter', do we still know what he is.[25]

DEXIPPUS: In reply to this we may say that what we have here is not a strict definition, but a rough description (*hupographê*), and anyhow, many definitions are presented through negations, when what is being asserted is familiar to us. Even so, in the case of where someone is defining the indifferent (*adiaphoron*), where 'good' and 'evil' are familiar concepts, one can say 'that which is neither good nor evil'. And
15 so here, it is for this reason that he made his preliminary statement about 'being in a subject and said of a subject', in order that by means of the negation of those attributes he might reveal substance in its strictest sense. And he did not

[24] This *aporia* and the reply to it is given by Simpl. in words closely akin to Dexippus (91,15-32), although he does not attribute it to anyone (*phasin*, 1,17 – perhaps Lucius and Nicostratus?). Plotinus does actually make this point at the end of 6.1.2,15-18. It is possible here that Simpl. is simply expanding Dexippus, but far more likely that they are both copying a common source, especially as Dexippus' version is distinctly elliptical, as if he were condensing a source. See on this, P. Henry, 'Trois apories orales ...', 236-40.

[25] That is to say, the essence, *ti estin*.

rest content with a simple negation, but he added an example, as if when saying that man was neither horse nor cow one were to add, 'like Socrates, for instance'.

[**9**. That the proprium of Substance <does not>[26] fit all cases.]

SELEUCUS: Once again, Plotinus raises the problem that the characteristic of being, while one and the same, alternately receptive of opposite qualities,[27] does not fit all instances of substance, either intelligible or physical. 20

DEXIPPUS: In response to this problem we would say that he has not presented here the proprium of every sort of substance. One can recognise this from the fact that he begins his preliminary description with 'is' (*esti*) (for he says 'Substance *is* that which is neither said of a subject nor in a subject'), and he immediately qualifies it by saying the 'so-called (*legomenê*) substance', that is to say what is called 'substance' in everyday parlance, and that to which one would attach this name; for philosophers use the word in one sense, and the general public in another. So it is of this popular sense of the word that the proprium is here presented, so that it would fit every case of the composite and its species and genera.[28] 25 30

[**10**. How it is that he now places sensible substance prior to all others, and calls it first, whereas elsewhere he places this after the incorporeal?]

SELEUCUS: But why does he here call primary substance the sensible, whereas elsewhere[29] it is ranked second after the incorporeal?[30]

[26] The text here omits the negative, but that does not square with the main text, so it should be inserted.

[27] *Cat*. 4a11, with the addition of *ana meros*. This *aporia* is raised by Plotinus at the very end of *Enn*. 6.1.2. It is not found in Simplicius.

[28] How this is a solution to the problem raised, I quite fail to see. Plotinus himself puts forward a solution in *Enn*. 6.3.4, but not on the basis of *Cat*. ch. 2.

[29] Perhaps such a passage as *Metaph*. 7.7, 1032b1-2, where he refers to Form as the *prôtê ousia*? Porphyry deals with this question at 91,13ff.

[30] The *aporia* is reproduced verbatim, along with the solution given here, by Simpl. 82,1-6. This is followed, however, by a report of Alexander's statement (not

45,1 DEXIPPUS: Because it is not his purpose here to speak about those, and because they are not called substances in common parlance.

[**11**. That he seems to be involved in inconsistency, since in the *Physics* he gives prior ranking to common items (*koina*) as primary, whereas here he ranks particulars (*ta kath' hekasta*) first.]

SELEUCUS: This question[31] also relates to the same enquiry, namely, why is it that in the *Physics*[32] he gives prior ranking to common items as primary, whereas here he ranks particulars first.

5 DEXIPPUS: This is because 'prior' and 'posterior' have two senses, either by nature or in relation to us. In relation to us, particular entities are first (for we come up against these first), but by nature common items are prior, for the particular is ranked under the universal (*ta katholou*). If, then, one were to begin from the nature of things (*phusis*), one will give prior ranking to simple entities, causes, things which 10 have their being in themselves, universals, immaterial entities, indivisibles and such like; but if their order is viewed from the perspective of their semantic relation to beings, then they will be ranked in the opposite direction.

[**12**. A refutation of Alexander's view of common items.]

SELEUCUS: But what could we say to those who dispute this very point, claiming that in fact universals are not prior in nature to particulars, but posterior to them?

---

necessarily an answer to the *aporia*) that the intelligible and separable Form is called 'indivisible substance', followed by a criticism of Alexander by Iamblichus. None of this is reflected by Dexippus, but that is hardly surprising.

[31] This *aporia* is found repeated virtually verbatim in Simpl. 82,22-35, probably from Iamblichus, whom he has quoted just above. From Simplicius one may gather that it was Alexander who first raised this problem, and attempted to answer it, and his solution is in turn criticised by Iamblichus, who produces the present solution.

[32] cf. 1.1, 184a23, where he says that we must advance (in philosophical enquiries) from *ta katholou* to *ta kath' hekasta* (though he is using *ta katholou* there rather in the sense of 'concrete whole' as opposed to its constituent elements); 1.7, 189b31, where he says, 'the natural order of exposition (*kata phusin*), as we have seen, is to start from *ta koina* and proceed to *ta peri hekaston idia*' (where, however, *ta koina* means 'general principles'); and 4.1, 200b24, where he says, 'for the treatment of *ta*

DEXIPPUS: Well, if we were to consider the question carefully, we would find that they actually take as agreed what they are disputing. For when they say that the universals are prior in nature to each thing taken individually, but on the other hand posterior in nature to all of them,[33] they are postulating that particulars are also prior by nature, and are proceeding invalidly in taking as a first principle that very thing that requires demonstration. Such attempts at proof (*epikheirêmata*) as the following are also superficial: if the common item (*koinon*) exists, it is necessary that an individual exists also (for individuals are comprehended in common items), but if an individual exists, it does not always follow that a common item exists, if at least a common item belongs to many (instances). For it is obvious that the particular can be an individual only if a common item is immanent in it and completes its essence, for a particular man is also Man. And it is also false to say that, when the common item is eliminated, the individual is not always eliminated also; for if the common item possesses its essence in its extension over all those things that fall under it, one who eliminates this eliminates straightway also the whole reality (*hupostasis*) of the common item[34] as well. It is on these lines that one must reply to Alexander and Boethus and the rest of the Peripatetic arguments in their attempts at explanation, and one should show, in interpreting the

15

20

25

idia must come after that of *ta koina*' (where the contrast is between properties peculiar to a particular science and properties common to all, such as motion). So the *aporia* is quite specious, and Iamblichus' solution, though making use of Aristotle's principle of procedure enunciated in *Phys.* 1.1, is quite beside the point.

[33] Reading *pantôn* for *pantôs*. On priority in nature, see *Metaph.* 5.11, 1019a2ff.; *Cat.* 13.

[34] The text of Dexippus reads *tou koinou*, but the parallel passage of Simplicius (82,30-5) reads *tou atomou*, 'the individual'; the argument would be that, if one abolishes the common term, e.g. 'man', the individual entity can no longer be categorised as e.g. a man. Steven Strange, however, contributes the following in defence of the text: 'The text *may* be sound. The idea anyway is that if all men besides Socrates were to disappear, there would no longer be a *common* nature of man; hence Socrates' existence or being does not depend on that of the common nature of man; it is not naturally prior to him. D's view seems to be that "the common nature of man" is just another way of talking about the nature of man, and this is part of what constitutes Socrates. So destroying its "whole reality" would destroy Socrates too. This perhaps shows that the text can be retained.' Cf. on this issue, Porph. 90,33ff., where Porphyry, we may note, is in agreement with Alexander. On Alexander's doctrine, see M. Tweedale, 'Alexander of Aphrodisias' views on universals', *Phronesis* 29, 1984, 279-303, who gives a most acute analysis of Alexander's doctrine, which is quite subtle, and has been misunderstood by all Neoplatonic commentators.

30 *Metaphysics*, that Aristotle takes common natures as prior in his theory of sensibles as well.

[**13**. That he should first state the reason why secondary substances are so called.]

SELEUCUS: Some people consider[35] that the order of topics is not correct here; for the logical order, they say, would have been, first, to state the reason why secondary substances are 46,1 'secondary', and then to make the comparison of secondary substances with one another.

DEXIPPUS: We say to that, that he has clearly presupposed this in the distinction he makes of things that are said of a subject and in the fact that that substance has been shown to 5 be most properly so called which is neither said of a subject nor in a subject; for from this it necessarily follows that those things are secondary substances which are said of a subject, but are not in a subject.

[**14**. That although he has not demonstrated that primary substance does not admit of more and less, yet he talks of species as being more or less.]

SELEUCUS: Further, Plotinus[36] raises an additional problem about the order of topics, that it would have observed logical order to demonstrate that primary substance was not more or 10 less substance, and then to talk of the more and less in relation to species.[37]

DEXIPPUS: It is possible to make the same answer here also, that he has already presupposed the fact that primary substance does not admit of more and less, when he declared

[35] This *aporia* and its solution are given virtually in the same words by Simpl. 89,23-30, also attributed only to 'some people'.

[36] Dexippus attributes this *aporia* to Plotinus, but it corresponds to nothing in *Enn.* 6.1. Simpl. reproduces this *aporia* and *lusis* almost verbatim (89,30-90,6), without the proper name (just *prosaporei*, which implies the same authorship as the previous *aporia*). Once again, Dexippus (and Simplicius) may be misattributing the *aporia*, but there is the possibility of an 'oral *aporia*' (Henry, however, does not discuss this passage).

[37] 'Forms' (*eidê*)? Both meanings seem to be envisaged. This *aporia* criticises the fact that only at 3b33 does Aristotle tell us that Substance does not admit of a more and less, whereas at 2b7 he talks of the species being more a substance than the genus.

that it was primarily and most of all and most properly substance; for all such things are definite, and do not have the indefiniteness of the more and less. For the more and less 15 proceed onwards always unceasingly, whereas 'most' (*malista*) is static in its supremacy, and 'primarily' in its being the beginning (*arkhê*), and 'most properly', in being the predominant characteristic of a given nature.[38]

[**15**. If no substance admits of the more and less, how is the species said to be more substance than the genus?]

SELEUCUS: But if no substance admits of the more and less (3b33), how is the species said to be more a substance than the genus (2b7)?[39]

DEXIPPUS: Because it is not in so far as they are substances 20 that they are involved in the more and less, but in virtue of their greater or lesser remove from sensible substance.

[**16**. If he takes the cause of sensible substance, he is taking that which is less substance.]

SELEUCUS: But if he is taking the cause of sensible substance[40] (for sense-objects are in process of becoming and not truly existent), is he not in saying this inappositely introducing an account of what is primary by nature?

DEXIPPUS: We are not here in search of what is primary by 25 nature, but of what is primary with reference to us and what are the primal elements in speech. If in fact we come up against species first, and then genera, and we signify species first, and secondarily genera, then it would be reasonable that species should be called 'more' substances than genera.

[**17**. That he has not defined generic substance.]

SELEUCUS: So much for that, then. But when he says 'in a 30

[38] This last phrase (from 'and primarily') is not found in Simplicius, being either omitted by him, or added (to Iamblichus) by Dexippus.

[39] cf. Porph. 97,7ff., and Simpl. 90,16-19.

[40] This seems to be a reference to a view of Alexander, cf. Simpl. 90,31ff. Porphyry also addresses the question, at 91,15ff., giving the same reply as Dexippus, whereas Simplicius gives a different one.

secondary sense genera and species of the primary substances are also called substance',[41] there has been, as it were, the division of a genus into species, so that Substance would be the common genus, with as species individual substance and secondary substance. Where, then, has he defined generic substance?

DEXIPPUS: My reply is that in giving a rough definition of 47,1 primary substance by saying that it is 'that which is neither said of a subject nor in a subject', and of secondary substance that it is 'not in a subject, but is said of a subject', he assumes as a characteristic common to both that of being 'not in a subject'. If, then, the characteristic of not being said of a 5 subject is peculiar to primary substance, that of not being in a subject will be proper to the common (genus of substance), and so this has not in fact been passed over.

[**18**. That he has not made a division of the substances.]

SELEUCUS: But if the division (*tomê*) is into species, and primary substance is not a species (for it is one and a unit numerically), whereas a species is many in number), how would what has been said constitute a division (*diairesis*) of Substance?

DEXIPPUS: We will say that being one numerically is the 10 common account of primary substance as a whole, which is particularly proper to the species, but common also is 'not being said of a subject nor being in a subject'. So the individual man or the individual horse as such is not a species, but, *qua* primary, all such substance would be a species.

[**19**. That the phrase 'It is clear from what has been said' has no clear reference.]

SELEUCUS: But having said, 'It is clear from what has been 15 said that, if something is said of a subject, both its name and its definition (*logos*) are necessarily predicated of the subject'

[41] This is not a literal quotation of Aristotle, however, who says (2a14-16): 'The species in which the things primarily called substances are, are called secondary substances, as also are the genera of these species.'

(2a19-21), he does not make clear where he has earlier said that 'if something is said of a subject, both its name and its definition are necessarily predicated of the subject'. For no clear reference emerges from what he has just said about Substance. For it was said that individuals are primary substances, while secondary are their species and the genera of these species, but this contributes nothing to the clarification of the statement that the definition and the name are predicated of the subject. 20

DEXIPPUS: Those who make this allegation are troubling themselves unnecessarily; for it is plainly implied in his previous statement (1b10): 'Whenever one thing is predicated of another as of a subject, all things said of what is predicated will be said of the subject also.' For it follows from this that that which is predicated of the subject must be true both in name and in definition of the subject.[42] 25

[**20**. That the study of differentiae is useful.]

SELEUCUS: Since he says of differentiae that they have in common with substances the characteristic of not being in a subject (3a21-22), from which he concludes that this is not a peculiar property of Substance, we must examine first of all what it is useful to know about differentiae.[43] 30

DEXIPPUS: It is agreed that the differentia is that which is of such a nature as to introduce distinctions into what is subsumed under the same genus and, in virtue of such distinguishing (*diakrisis*), in relation to the genus it constitutes its division (*diairesis*). It would be more proper to say,[44] however, that one should connect the differentia with the species, since often they are employed as equivalent to the species. For all the differentiae collectively would be said of the genus, but each individually would be said of items that 48,1

[42] The previous three sections (17-19) are not reflected either in Porphyry or in Simplicius. They are perhaps Dexippus' own contribution. Connexion with Simplicius resumes with section 20.

[43] This passage, down to 48,19, is to be found closely paralleled in Simplicius, 97,24-98, though in a rather more expanded form, and is more loosely paralleled in Porphyry 95,10-96,2.

[44] Simplicius here (97,28) specifies that this more correct definition is the work of Boethus. It seems here as if Dexippus is abridging his source, since Simplicius would have no independent basis for expanding it.

5 fall under it, of which it is said overall, but not at all of the genus; for each individually is not formative of the genus.

The differentia differs from the species as whole does from part. For a differentia is a quality constitutive of essence (*ousia*) and it is not on the one hand in a subject, because it contributes to the essence of those things which it specifies, but it is not on the other hand a substance, because it does not contribute to the existence of a thing, but to its being of a 10 certain quality. We must understand 'of a certain quality' not as we say, e.g. of white; for this is separable, whereas the differentia is not separable, except in so far as it might perish along with its subject. But if someone were to say that the essential accident[45] (*ousiôdes sumbebêkos*) also is not separable without perishing, we will reply to him by signifying the complete difference between the essential accident and the differentia. For one can observe degrees of 15 intensity in respect of essential accidents, as for instance, in the case of an Ethiopian moving to other climes there would come about a lessening of the blackness of his skin, and the whiteness of milk is lesser by comparison with that of snow; whereas in the case of a differentia, one would not observe any variation of more and less; for one biped or footed thing is not more or less so than another.

[**21**. A problem about where one should class the differentia.]

20 SELEUCUS:[46] But if the differentia is neither a substance nor an accident (*sumbebêkos*), and there is no third thing other than substance and accident (for all existent things are either in a subject or not in a subject,[47] for they are either

[45] I adopt this translation, despite its apparent absurdity, since that is what the Greek *says*. Simplicius exhibits a slight variation in terminology here, instead of an *ousiôdes sumbebêkos* talking simply of an 'inseparable accident' (*akhôriston sumbebêkos*) as does Porphyry before him, at *Isag*. 8,134; 12,26.

[46] This section is also to be found in Simplicius, following on directly from the previous one (98,18-35), and so probably Iamblichean. Cf. Porphyry's discussion of the question, 95,10ff.

[47] The mss. of Dexippus here actually read 'or said of a subject' (*kath' hupokeimenou*) and Busse has corrected the text from the parallel passage in Simplicius. He is surely right. The error in the Dexippus mss. may have resulted from the *oukh* falling out, and an 'intelligent' scribe emending the text.

substances or accidents); so if the differentia is not to be classed with either of these, where are we to find a place for it?

DEXIPPUS: There are many possible solutions to this, which I will communicate to you in a little while, but first I want to propound an axiom, as it were, on the basis of which we will be able to judge the problems and their solutions. It is the view of those who are experts in the investigation of reality[48] that the passage, in nature, from one genus to another is inexpressible, because the intermediate stages are imperceptible to us, as for instance potentiality in the case of substances – I mean, for example, in the case of men or other animals, when viewed in the seminal state, the thing is neither yet a substance (for it is still incomplete), nor is there any actuality attached to this potentiality (for actuality is something complete); the remaining possibility is that there is something else intermediate between, for instance, the actual man and the potential man. It is the same story, then, also with the differentia; for it too will hold an intermediate rank between quality and substance, because it contributes both to something's being and to its being of a certain sort (*poion*), and so it will not be *in* a subject. For this reason one should add to the definition of what is in a subject the qualification 'contributing nothing to the substance of the subject'; for both the species and the differentia do contribute something.[49] 25 30 49,1 5

So, after these preliminaries, let us take up the solutions of the philosophers of old on the question of the differentia.[50] One position would be that which postulates that the differentia is a quality, but a quality which is essential (*ousiôdes*) and constitutive of substance. Another would be that which takes the differentia as being median between quality and substance, providing a sort of common bond for substances with their essential attributes, and conversely for attributes with substances. For Nature does not pass directly 10 15

[48] A common way of referring to the Stoics, though perhaps a curious way for a Platonist to do so. Presumably a reference to the *sorites* argument (*SVF* 2.274-7).

[49] This argument is attributed explicitly to Iamblichus by Simplicius, 99,4-9, so it is probable that Dexippus has taken this whole line of argument from Iamblichus.

[50] The following passage, to the end of the paragraph (49,25) is reproduced verbatim in Simpl. 98,19-35, though without the attribution to *palaioi philosophoi*. The paragraph is very probably Iamblichean (see previous note), though the first view presented is actually that of Porphyry (95,22ff.), and the second (49,20ff.) that of Plotinus (*Enn.* 6.3.5,25-7).

from opposites to opposites, as for instance from animals to plants, but she has also contrived a type of life median between the two, in the shape of the class of 'zoophytes',[51] which links the two extremes, and completes and binds each to the other. And in this case also, between the different genera, as it were, of substance and accident there will be a mediating entity, according to some as partaking of both these, while according to others as distinct from both of them.
20 But another view might be put forward,[52] that the differentia is not only constitutive of the substance, but also a part of it, either when viewed as including the subject, being a part of the substance in its formal aspect, or as changing along with the relationships (*skheseis*) of the subject, as being in various states and dispositions relative to the subject.[53]

[**22**. A problem arising from the fact that differentiae are not 'said of' a subject.]

25 SELEUCUS:[54] But it seems to me in no way true to say that the differentia is predicated of a subject; for it is neither possible for the name nor the definition (*logos*) of the differentia to be predicated of the subject. This is obvious from the following example: 'footed'[55] means on the one hand the state itself, 'footedness', as it were,[56] but has as another
30 meaning that which partakes in this state, which is not the differentia, but that which owes its disposition to it. This being the case, let us leave aside 'footed' for the moment (for
50,1 the differentia is not the same as that which derives its disposition from it), and take the actual differentia itself,

[51] That is 'animate plants'. The concept of this kind of intermediate class of beings (e.g. sea-anemones, corals and sponges) may go back to Aristotle, but the actual word does not seem to be attested before Dexippus (= Iamblichus). The occurrence in the text of Sextus Empiricus, *PH* 1.41 is a false reading.

[52] This phrase (*allê d'an genoito doxa*) is characteristic of Iamblichus' way of introducing his own opinion, cf. *DA* 372,15 Stob., *genoito de k'an allê doxa ouk apoblêtos*.

[53] If this is the sense of *ekhein pôs … kai diatithesthai*.

[54] Once again, this problem and solution are given in substance, though not in identical words, by Simpl. 100,13-101,11, the passage being attributed explicitly to Iamblichus (101,12). This is all a criticism of *Cat.* 3a21.28.

[55] In Greek, *to pezon* is indeed ambiguous, as being the abstract quality and the qualified thing.

[56] *Pezotês* would seem to be an Iamblichean coinage (also found at Simpl. 100,20); hence the cautious *hoion*.

'footedness'. Is it possible to predicate the name of the differentia of the subject, so as to say that the subject *was* 'footedness', or yet its definition, so as to say that the subject is 'a state (*hexis*) arising from which participants in it become footed'? One cannot say this. It is invalid, then, to say that the 5 name and the definition of the differentia can be predicated of the subject. So the differentia would not even be said synonymously of subjects.[57] Therefore it is not 'said of a subject', if it is the case that what are said of a subject are predicated synonymously of those individuals falling under them.[58]

DEXIPPUS: To this problem we might make the following 10 reply. I am not prepared to abandon, as they wish to do, that which possesses footedness on the grounds that this is not a differentia, but to assert that on the contrary 'footedness' is, as they maintain. For they, in a purely conceptual (*epinoiâi*) distinction of footedness, are not prepared to accept that the differentia is that which is taken along with the subject. But it is one thing to distinguish something conceptually, and another to take something as belonging together with (*sunuparkhon*) its subject. Therefore, when Aristotle, in 15 connexion with his definitional (*horistikos*) account, says here that being predicated synonymously is characteristic of differentiae, he is talking about things taken together with their matter, and we understand him in that sense; for 'footed' comprises also 'footedness', but 'footedness', being a mere concept, does not comprehend footed. One should not therefore separate the account in terms of the underlying 20 form[59] from the definitional account; for if the definitional account reveals the essence (*to ti ên einai*), and this is the form, then there will be no difference between the account in terms of the definition, since it would be paradoxical that something should relate to something in the area of essence,

[57] cf. *Cat.* 3a33: 'It is a characteristic of substances and differentiae that all things called from them are so called synonymously.'

[58] I append Ackrill's comments on this problem (*Comm.* p. 86): 'Aristotle, indeed, positively claims that the definition as well as the name of a differentia is predicable of the substance falling under it, but this too seems very strange. In a definition *per genus et differentiam* the differentia is commonly expressed by an adjective (or other non-substantive).' Plotinus raises this problem at *Enn.* 6.3.5,27-9.

[59] Or 'the form as subject' (*to hupokeimenon eidos*)?

but that the name and the account should not be true of it as a subject.

25 This is one possible reply, then; but we can state another solution not to be despised.[60] Since that which most of all creates the form[61] is the differentia, he has declared it to be true in the statement that the account of the subject is completed by this; so it is 'said of the subject', as they[62] would wish, because it is completive of the account related to the subject. 'For if I say "footedness",' he says,[63] 'I mean "motion on 30 land by means of legs", and "motion on land by means of legs" indicates both the footed and what is moved in this way.'

So much for that, then. But I am surprised at the Stoics making a distinction between states (*hexeis*) and subjects of states (*hekta*),[64] for although they do not accept that they are incorporeal as such,[65] when they feel the need to play games with words (*ereskhêlein*), they resort to such distinctions. But if qualities, for instance, do not exist without qualified things, then neither can states be contemplated without subjects of 51,1 states nor differentiae without a subject, but the subjects will be constituted by their own definitional account.

[**23**. How is that having said 'it seems to pertain to every substance that it signifies a certain this', (3b10) he then goes on to say 'however, this is not really true' (3b15)?]

SELEUCUS: 'But does he come suddenly to change his position? For having said that 'it seems to pertain to every

[60] This is a characteristically Iamblichean turn of phrase, cf. n. 50 above.

[61] Or 'the species'? The problem is that Aristotle was not really aware of any such distinction. Cf. n. 36 above.

[62] sc. those who have raised the *aporia*.

[63] sc. Aristotle.

[64] Only Dexippus (here = *SVF* 2.461) and Simplicius (214,24ff. = *SVF* 2.391) credit the Stoics with this distinction (both doubtless dependent on Iamblichus and/or Porphyry). Simplicius tells us that the Stoics gave the name of *hekta* to qualities (*poiotêtes*), though one would expect rather the equivalence to be with qualified things (*poia*). But Simplicius goes on to make a distinction between *hekta* as only applicable to unified things (*hênômena*), whereas *poia* may characterise joined-up things, like a ship, or even discrete things, like an army, so he may know what he is talking about. 'Subjects of states' for *hekta* may be inaccurate, but I am at a loss for better. Long and Sedley propose 'havable', which is more accurate, if one can stomach it.

[65] I understand an *auta* here, agreeing with *asômata* (incorporeal), referring to *hekta*, since it is hardly possible for Dexippus to assert that the Stoics did not believe in incorporeals *at all*.

substance that it signifies a certain this',[66] he then goes on to say, 'however, this is not really true'.

DEXIPPUS: Our reply to them[67] is that the first opinion was not his but the view of others; for certain people hold this opinion, who claim that genera and species subsist (*huphestanai*).[68] The second is his own opinion, so, having first set out the view of others, he later added what he himself approved. 5

SELEUCUS:[69] But how in that case do we say that the individual man is a man and an animal and rational and mortal, if all these things are not 'thises'? 10

DEXIPPUS: Because it is when species are linked (*katatakhthenta*) to individual substances that we speak of them as 'thises', but the subject of discussion here was species thought of as on their own.

SELEUCUS: But in what respect do we speak of an individual substance as a 'this', in respect of its specific form (*eidos*), or its matter, or as a compound of the two? 15

DEXIPPUS: We will speak of it as such in all respects, but in respect of its matter in so far as it acts as a substratum and comes to actuality in the receipt of form, while in respect of its form in so far as it is definite and numerically one. And if someone objects that matter, being indefinite, is in no way a 'this', we will remind him that the present context concerns not matter in its absolute state, but only matter in relation to form. 20

[24. If there is no contrary to Substance (3b24), how comes it that rational animal is contrary to irrational animal?]

SELEUCUS:[70] How is 'rational animal' not contrary to 'irrational animal'?

[66] This is an inaccurate quotation of 3b10, contaminated by reminiscence of 3a33-4 and 3b33.

[67] sc. those raising the *aporia*. Since they have not been mentioned, we seem to have evidence here of summarising by Dexippus. The *aporia* does not appear in Simplicius, though he deals with the problem, 102,13-104,18.

[68] Presumably the Stoic contrast between *huphestanai* and *huparkhein* is not being envisaged here. Those referred to would seem to be Platonists, though this is a rather offhand way to refer to one's own school.

[69] This passage, down to section 24, is reproduced verbatim in Simplicius, 104,19-30, with two small additions, one mentioning Plato, the other Boethus.

[70] This passage (to 52,4) is reproduced by Simplicius, 106,28-107,4, virtually

DEXIPPUS: Because[71] a differentia present in any given thing may be opposite to another differentia, but a whole is
25 not opposite to another whole. The reason for this is as follows: that which acts as a substratum for opposites is not itself an opposite; for if it were taken over by one of a pair of opposites, it would not be able to be suitably placed to receive the other opposite, as for instance soul, body, individual substance or secondary substance, even if they receive opposites, will not be in themselves opposites. But not even if
30 one were to divide something into opposites, as for instance 'animal', would it even so be an opposite; that embraces the
52,1 division into opposites, so that none of these things will be an opposite.

One might recognise this from the definition of opposites; for we define them as what are at extremes of separation from one another, and, as is being stated here, belong to (*huparkhei*)[72] the same substance (as in this case, animal).

[**25**. How is it that, if there is nothing contrary to Substance, Aristotle talks of fire and water and other elements as opposites?]

SELEUCUS:[73] But how is it that Aristotle talks of fire as
5 'opposite' to water, and air to earth?[74]

DEXIPPUS: Obviously, we shall say, by reason of their specific differentiae, which are not substances; for cold and hot and dry and wet are opposite to each other, but the substances as wholes do not have a relationship of opposition to the other substances as wholes, seeing as they subsist
10 together in the same matter.

---

verbatim, and presented explicitly by him as a quotation from Iamblichus, who is countering an *aporia* by *tines*. Simplicius also quotes the passage at greater length in his *de Caelo* commentary (169,2-27). This passage is most valuable for studying the complex relationship between Iamblichus, Dexippus and Simplicius.

[71] Emending *hoti hê en* to *hoti men* at the beginning of the sentence.

[72] Simplicius' text in both *in Cat.* and *in de Caelo* reads here *sunuparkhei*, so that is presumably what Iamblichus said, whatever about Dexippus.

[73] The *in de Caelo* text shows that this section was continuous with the previous one in Iamblichus' commentary.

[74] As e.g. at *GC* 2.3, 330a30ff.

[**26**. That Man can be opposite to Horse, or a particular man to a particular man.]

SELEUCUS: But what if one were to say that a man is in a way opposite to a horse, or a particular man to a particular (horse)?[75]

DEXIPPUS: To him we would reply that there is no way that any particular horse or ox or any other of those entities equally on the level of primary substance is any more or less opposite to any particular man. But if it is impossible for there 15 to be opposition between indefinite things and any definite thing (for one thing must necessarily be defined as opposite to one thing), then it would not be correct to take these antitheses in this sense either.

[**27**. How is it that, if there is nothing contrary to Substance, he says that Form is contrary to Privation in the Physics, and Matter to Form?]

SELEUCUS: But how,[76] if he says that Form is contrary to Privation (*sterêsis*) in the *Physics*,[77] is it correct for him to declare now that nothing is contrary to Substance?

DEXIPPUS: Just because he was accustomed in a more 20 general way to call 'contraries' things that are opposed as form to privation, as when in the *Physics* he calls possession (*hexis*) of a form and privation 'contraries', it should not disturb us at all that, whereas in the *Categories* he has said that one substance is not contrary to another, in the *Physics* he seems to go back on himself and accept contrariety

[75] It seems possible that the compiler of the headings has misunderstood this and that Dexippus means to contrast particular man with particular *horse*. This *aporia*, to which I confess I cannot attach much sense, is not to be found in Simplicius, and is thus probably to be attributed to Dexippus himself, since in Simplicius the next *aporia* is found following immediately from the one before this, probably reflecting the situation in Iamblichus.

[76] This *aporia* and *lusis* are found in Simplicius 107,30-108,21, but in considerably altered form. It is not clear who has strayed further from Iamblichus. Only at 52,28-30 (= Simpl. 108,17-19) is there verbal coincidence.

[77] 1.190b27ff. In fact, in this passage he says that a given quality and its privation are '*in a way enantia*, and *in a way* not', but that is only because they cannot act or be acted upon directly by each other.

between them. For he uses the general term 'contrariety' also to describe all other types of opposition.

25 Now this solution is not to be rejected, it seems to me, but there is another way of solving the problem that is both more elegant and more appropriate to Aristotle, and that is that if [78] form is opposed to privation not as substance to substance, but as being to not-being,[79] or first principle (*arkhê*) to last element and good to evil, it will possess complete contrariety

30 by reason of total distancing or descent (*huphesis*). For if we make this division, what we see is not a contrariety of

53,1 substances, but a different distinction of contrariety in the area of first principles will make its appearance, which does not involve entities of the same genus; for neither do both contain contraries in existence (*huparxis*), nor are they genera which are contained (by others), but the one is existent, the other non-existent.

[**28**. That it has been demonstrated that there is nothing contrary to Substance at the level of individuals, but it has not been demonstrated for Substance in general.]

5 SELEUCUS:[80] The thesis that there is nothing contrary to Substance has been demonstrated by induction for individual substances, but it has not been demonstrated for Substance in general.

DEXIPPUS: The propounder of this difficulty is ignorant of the method of Division and its use. For since it is not possible to construct a demonstration of the most generic entities from prior and more basic causes, but one must rather first divide

10 with perfect divisions all the entities ranged under them and thus reveal their commonality (*koinotês*) lest one miss something,[81] as for instance when we divide 'living thing' into

[78] It seems to me that one must insert an *ei* here, in order to preserve the syntax.

[79] Dexippus' mss. have *tôi eidei* here, to which I can attach no meaning. It is omitted in the corresponding passage of Simplicius (108,17), and that seems the best solution.

[80] This *aporia* is taken verbatim from Plotinus, not from *Enn.* 6.1, but actually from 1.8.6,28-31, the treatise on the nature and origin of evil. This whole section is reproduced almost verbatim in Simplicius 108,21-109,4, where the quotation from Plotinus is acknowledged (108,22). Once again, Iamblichus is presumably the common source.

[81] Emending *hêi parallattei*, on the basis of the parallel text of Simpl. 108,30, to *mê parallattêi ti*, as Busse suggests.

rational and irrational, or footed and footless, or biped and quadruped, or such other divisions as present themselves, and then, when we find that none of these differentiae[82] goes beyond the area of 'animal', conclude the identity of the genus not from induction but from perfect division. So then, even as in other cases division preserves the common element in the 15 various sub-divisions of a concept, so it is shown to be the case now with substances. For if all substances are now divided perfectly into primary and secondary substances, and neither of these admit of contraries (for the division, being perfect, takes in all that falls under substance) then not only a particular substance has no particular substance contrary to it, but there is no contrary to Substance in general. 20

One might also deal with the problem as follows: when one takes one of those qualities which are constitutive of substance, this is shown to be true simultaneously in the case of each and of all, as for instance 'rational' in the case of each man and of Man in general, and so in the case of Substance, if it is true of Substance not to have any relation of contrariness, this will be true not just of some particular substances, but of 25 all.

[**29**. How, after he has said that the species is more substance than the genus, can he now say that Substance 'does not admit of the more and less'?]

SELEUCUS: Again, there seems to be the following difficulty: how is it that in an earlier passage (2b7) the substance of species has been determined to be more of a substance than the substance of genus, while now (3b33), there is said to be no 'more and less' in Substance?

DEXIPPUS: The explanation is that he has not made these two statements in the same sense, but the latter refers to an essential, the former to an accidental characteristic; for to say 30 that the species is 'more' than the genus is not to say that it admits of 'more' *qua* substance or as being what it is, but just in so far as the one approaches more nearly individual substance, whereas the other, subsisting to a lesser degree, is

[82] Inserting *diaphoran* after *heuriskontes* from the parallel passage of Simpl. (108,31), as Busse suggests.

54,1 further removed from it. In the present context, on the other hand, he is talking of each substance as being what it is when he says that it is not said to be 'more or less'. So there is no contradiction between the present passage and what was said earlier.[83]

[**30**. That he who is more rational is more a man.]

SELEUCUS: Still, this problem is worth raising, how is it that he who is more rational is not more a man?

DEXIPPUS: Because man does not have his being in the degree (*epitasis*) of his rationality, but in his form (*eidos*). But this is always static, while it is the activity of a man according to the differential of his increasing rationality which brings it about that a man is more rational, not his humanity; <for it is by reason of his humanity>[84] and his capacity for being rational that he is endowed with existence as a man. So this particular man is the same, but his activity as a rational being takes on greater or lesser intensity.

I see some merit in this solution, then; but if one is to give a reply more deeply in accord with Aristotelian doctrine, we might provide a solution to the problem raised on a more theoretical level. It is a fact of nature that along with each form there goes a quality, distinct from the form, but dependent upon it; and the form, being a part of the individual composite substance, is constitutive of it and never admits of more and less, but the qualities which go along with it, whether it be rationality, or heat, or dryness, do admit of degrees, and so this apparent variation of degree is not a function of the form but of the quality, so that when the good man (*spoudaios*) is said to be more of a man, this increase in degree is true of him not *qua* man, but *qua* man as being in a certain condition. But this is a matter not of substance, but of quality, so that we are right to accept the doctrine that one

[83] Aristotle himself, of course, is quite well aware that he has uttered an apparent contradiction, and proceeds to clarify his meaning immediately afterwards (3b34ff.), which makes this *aporia* rather superfluous. Simplicius reports it, in more or less the same words as Dexippus (111,10-18), not as a real difficulty, but as a point requiring clarification.

[84] Inserting *kata gar tên anthrôpotêta* after *ou kata ton anthrôpotêta*, as supplied by Felicianus in his Latin translation. It will have fallen out by haplography.

substance is not more or less so than another.[85]

[31. How is Matter said to be 'more and less'?]

SELEUCUS: How, then, is it that Matter admits of the more and less, and reflects images of contraries in itself, in the shape of the great and small, and deficiency and excess? And how is it that in everything that it proclaims itself to have it is deceptive, that is, if it appears great, it is small, and if it appears more, less? 25

DEXIPPUS: We will reply to that that these characteristics come to Matter from its relation (*skhesis*) with Form, since it takes on all these mutations according to its relation to the form, in relation to which it subsists in one position or another. For according as the form is more dominant in it, it appears more actualised (*energos*),[86] while according as it is less so, it manifests itself correspondingly more dimly in the matter. In itself, however, Matter possesses no differentiation in any of these respects; for it is potentially all things equally, and exhibits equal reflectivity (*emphasis*) and willingness to receive their substance, in relation to all existent things.[87] 30

[32. What is the reason for Substance not admitting of more and less, but of qualities doing so?]

SELEUCUS: Again,[88] this too seems worth enquiring into, why substances do not possess more and less, while qualities admit of this? 55,1

DEXIPPUS: This question, too, we will be able to solve by

[85] This *aporia* has a complex relationship to Simplicius, who covers the same ground at 112,15-31. Only 15-22, however, corresponds to Dexippus and that to his second, more 'theoretical' solution, which is probably that of Iamblichus. The first solution could be that of Porphyry (perhaps as reported by Iamblichus); it does not correspond very closely to that found in Porphyry's short commentary (97,7-22), but it is doctrinally compatible with it. As for Simplicius, he prefers a formulation of his own (*mêpote oun* 112,27-31).

[86] Unless one should actually read *enargoteron* for *energoteron*, as being a more obvious contrast to *amudroteron*, 'more dimly'.

[87] This *aporia* is closely paralleled in Simpl. 112,32-113,5 except that Simpl. makes a point of adding (or Dexippus chooses to omit?) that Matter is itself a sort of Substance (*ousia tis ousa kai autê*).

[88] This *aporia* is not found in Simplicius. It adds little to what has gone before, and is probably a contribution of Dexippus himself.

applying the same line of reasoning, if we explain that substances, by reason of being 'in themselves', remain in the
5 sphere of the same definition, while qualities, by reason of finding themselves 'in others', suffer change along with them, and so it is natural that substances should not admit of degrees of intensity either from themselves or from others, whereas qualities range in intensity according to the substrata in which they come to be.

[**33**. Why did he add 'it seems', when giving the proprium of Substance?]

SELEUCUS: How can we assert that many people are
10 unreasonable to blame him for being indecisive and not stating his opinion clearly? For in giving the proprium (*idion*) of substance, he says: 'There *seems* to be most of all proper to it the characteristic of, while being numerically one and the same, being receptive of contraries' (4a10-11).

DEXIPPUS: My reply is that 'proprium' is said in many ways, but its strict sense is that which is true of *all* and *only* a given subject. 'It seems' is therefore attached to it in the present context since he is indicating that it is not possible for this to
15 be the proprium of substance in general and in the primary sense, in the way that a real proprium naturally would be, and that it is said to be a proprium in such a way as not to be true of all substance, both in general and in particular, but that such (a definition) is a proprium in the area of substance alone, and not in any other of the categories. It is in this way, after all, that fire is said to be the most tenuous (element), not because all fire is to an equally high degree tenuous, but
20 because the quality of being most tenuous resides in fire. And so in this case since, if there is any individual which is receptive of contraries, it falls under substance, for this reason it *seems* to be proper also to substance in general to be receptive of contraries.

We will realise this more clearly if we convert the terms (*antistrophê*) of the proposition; for if on the one hand something is a substance, there must be some individual falling under it which while being one and the same is
25 receptive alternately of contraries; while if on the other hand

something while being one and the same receives contraries, it is a substance. So then it will be the proprium as being said of none of the other (categories), but only of substance, and, deriving from this, of the individuals falling under it.[89]

[**34**. Why did he add 'numerically one' to (the definition of) Substance?]

SELEUCUS: But one cannot surely dispute that by adding 'numerically one' he has restricted this to being the proprium of individual substances only; for this is not a proprium of genus or species. 30

DEXIPPUS:[90] It is not difficult, in fact, to counter this objection, by pointing out that one type of knowledge is that which from the description of the part simultaneously reveals the whole; and so here too, if this is the proprium of the individual, the same can also apply to the genus; for since it is by virtue of the individual substance and by being present in it that secondary substances are said to be receptive of contraries, for this reason he gives this as the definitive proprium of them as well, showing that it is not the proprium of the simple concept 'man' that he is now defining, nor yet of the common notion of animal conceived on its own, but of that which is 'in relation' (*katatetagmenon*)[91] and actualised and in a particular 'this'; for it is man as viewed in individual instances, and animal as already related to particular (animals) that admits of such a proprium. 56,1 5

[**35**. That in wishing to establish that the property of being alternately receptive of contraries is true of all substances, we inadvertently establish a worse consequence, that it is not true of it alone.]

SELEUCUS: But if someone, picking on these proofs and presenting to us instead a worse consequence, says, 'Well then, what has been stated is not the proprium only of 10

[89] This final section is represented, rather loosely, by Simpl. 113,27-31, but what goes before finds no echo in Simpl. With Dexippus' reply to this *aporia*, cf. that of Porphyry 98,35ff.

[90] With this reply, cf. Porph. 99,19ff.

[91] For use of this term, see Book 1, section 26 above.

substance – for states and dispositions are also receptive of contraries – what shall we say to him? For a state and a disposition and an action is said to be bad and good, and a motion comes to be fast and slow, so that the characteristic of being receptive of contraries pertains not only to substances, but to many other things as well.'[92]

15 DEXIPPUS: To those who say this one may make reply on the basis of what has already been said, for it is not as considered in themselves, distinct from individuals, that we say that the genus and species of substances are receptive alternately of contraries, but as viewed as present to individuals and co-subsisting with them. So being alternately receptive of 20 contraries is to be viewed differently in the case of substances and in the case of motion and action and disposition and such like. For a substance, while being one and the same (like Socrates, for example), is receptive alternately of contraries, so that also the species Man and the genus Animal will be receptive of the contraries observable alternately in him, because of the fact that Socrates is a man and an animal.

But in the case of the other things mentioned the situation 25 is not the same; for this particular action,[93] say that of Aristides, is noble and this action, say that of Eurybatos, is evil,[94] and similarly the state of mind (*hexis*) underpinning the action, whether noble or evil. So they are different in each case, according as they become noble or evil, changing along with the one or the other of the contraries, not, like the particular substances, abiding the change of opposites. For 30 example even as the number four, if it loses its fourness, is no 57,1 longer even, so also the noble action, of Aristides for instance,

[92] This is presented by Simplicius, in much the same terms, as an anonymous *aporia* (*phasin*) – probably Lucius and Nicostratus – at 114,5-8. The *lusis* following here, however, is only very loosely represented in Simplicius (114,8-20).

[93] The peculiar inclusion of *actions* as Substances is, of course, provoked by Aristotle himself, cf. 4a15. Cf. Porph. 98,29-32.

[94] Aristides is well enough known, but Eurybatos may not be. He was in fact a proverbial traitor, who took money from King Croesus to raise troops in Greece, but betrayed him to the Persian King Cyrus instead. He is mentioned by Plato *Protag.* 327D (whence, doubtless, he becomes an exemplum in the Platonist tradition), and by Aeschines *C. Ctes.* 137, in company with a certain Phrynondas. For the full story, see Diodorus Siculus 9, fr. 32. What exact action of Aristides' is envisaged here is not clear. Perhaps that described by Plutarch in his *Life of Aristides* ch. 25, the removal of the treasury of the Confederacy from Delos to Athens, contrary to oaths he had sworn. The example would be most apposite if both actions were of the same sort.

if it is deprived of being noble and just, can no longer be said to be the action of Aristides; or again, the action of Eurybatos, if it has no longer the characteristic of being evil, can no longer be said to be the action of Eurybatos. For if the noble action, say, of Aristides, could have become also evil, then through losing its nobility it would yet have remained an action;[95] but since, if it was deprived of the essential characteristic of being a noble action, it would no longer even be denominated as such, it is perfectly plain that it is instantiated only along with nobility, and the evil action in turn with evil. Substances, however, remain the same even when their contraries change.

[**36**. That a father receives contraries, while being a substance; for he gets ill and gets well.]

SELEUCUS: But how is it that a father[96] is now well, now ill?

DEXIPPUS: We will say, because he is a man; so such a predication as this comes to be only accidentally. But the proprium of substance is not something accidental, but essential (*kath' heauto*) and not in relation to something else, but in virtue of a change taking place in itself.[97]

[**37**. That this proprium is not relevant to sempiternal[98] substances.]

SELEUCUS: But of what contrary will the sun be receptive, seeing as it is always stable identically in the same form? For there is nothing contrary to its nature. And the heaven would not ever be at rest, but rest is the contrary of motion. Furthermore, fire, among things that are destructible, is receptive of heat, but not of cold, and snow is receptive of cold, but not of heat.[99]

[95] I do not see why Busse rejects Spengel's proposed addition of *an* after *êdunato* here. The condition is surely counterfactual.

[96] 'Father' being a relational term.

[97] No trace of this in Simplicius, so perhaps a contribution of Dexippus.

[98] Taking this to be the meaning of *aidios*. Note that the chapter heading only represents the first part of Seleucus' question. The *aporia* relates to the heavenly bodies. Cf. Porph. 98,36-99,3.

[99] Here we have a most interesting situation, where Dexippus' text is matched virtually verbatim by Porphyry on the one side (98,36-100,9) and Simplicius on the

DEXIPPUS: In reply to this we would say that here he is laying down the proprium of all substance that is capable of coming to be subject to change, but not of that which in the realm of Being
20 embraces the unchangeability of the Forms. Further he says that it is receptive of contraries, not that it is instantiated (*ousiôsthai*) in contraries.[100] Fire, in fact, does not *receive* heat; heat is present in its substance. Nothing is receptive of itself but rather of something external, as for instance water receives heat, which is an acquired (*epiktêtos*) quality for it, but not
25 wetness (for this is connatural with it), and earth receives wetness, as being a quality external to it, but in no way does it receive dryness, for that is coexistent with its nature.

The heavenly bodies, again, have circular motion as part of their being, and so they would not admit of any receptivity of its contrary. Furthermore, such characteristics as these are not actually qualities, but essential differentiae which are consti-
30 tutive of a thing's substance, so that while the thing remains in existence it would not divest itself of what pertains to it substantially. So it is only those things which possess one contrary or another not by nature or inseparably that are 'receptive of contraries'.

[**38.** If it is the case that, even as the genus and the species do not in themselves receive contraries, so neither do states and dispositions, but only the sensible objects subject to them, then the characteristic of being receptive alternately of contraries would not be the proprium of substance. For what is seen to be true of a generic state or disposition is also true of generic substances, and thus not being alternately receptive of contraries will be common also to other things.]

---

other (114,21-115,10), betokening either direct borrowing from Porphyry, or wholesale adoption of Porphyry first by Iamblichus and then by Dexippus. This is an unusual passage in Porphyry's commentary, being a much longer piece of continuous exposition than the average. It may be taken directly from his big commentary, in which case it would give us an insight into the true relationship between the document and Iamblichus' commentary, as discussed by Simplicius in his preface.

[100] This distinction between 'acquired' and 'essential' attributes is reminiscent of that produced in section 20 above, in connection with the differentia.

SELEUCUS: Let us go back here to a former problem.[101] Even as genera and species of their own essence were not equipped to receive contraries, but only in virtue of the individual 58,1 substances falling under them, so also do neither states nor dispositions nor motions receive contraries in themselves, but only in the sensible objects subject to them.

DEXIPPUS: Our position is that it pertains to substance to be receptive of contraries, because it is a substratum in every 5 case for everything and everything belongs to it and in it, so that contraries also exist in relation to it, whereas the characteristic of being receptive of them is proper to none of the other (categories). For accidents, after all, are[102] not sufficient in themselves for their own substantiation, but they need some other base for their existence, and do not start off as a subject for anything. So how would things be receptive of contraries which do not even maintain their existence? For the subject, as for instance a body, becomes white and black 10 while remaining body, but colour does not, nor when from being white it becomes black, is the white present to receive the black, but the one no longer exists, but it departs, and the other comes to be. And Animal and Man, since they stay in place, receive the contraries, inasmuch as they are substances in the particular man, but colour does not stay in place, and so 15 does not receive; for when white departs, the colour departs with it and will no longer exist, and simultaneously with the coming into being of the blackness there comes into being the colour associated with it, or if not this, at any rate it exists when the other exists, but certainly the colour does not stay around to receive things passing out of it and into it.[103]

[**39**. A problem raised by Plotinus, to the effect that being alternately receptive of contraries is like an accident of substance.]

[101] Referring back to section 35. This *aporia* is reproduced virtually verbatim by Simplicius, 115,11-23. Presumably Iamblichean.

[102] Reading *exarkei* for *exêrkei* with the mss. of Simpl. I see no reason for the imperfect tense.

[103] The *eisionta kai exionta* here is presumably an echo of *Tim.* 50C4-5,where the reference is to the copies of the Forms, entering and exiting from the Receptacle.

20 SELEUCUS: Again, Plotinus raises the problem[104] that the alternate reception of contraries is an accident (*sumbebêkos*) of substance, so that he has not stated what Substance is, nor has Aristotle provided an account of what always properly (*idiôs*) belongs to it, but has merely indicated what is a concomitant of it another way.

DEXIPPUS: We shall say, then, to the man who raises this problem that he is demanding something more than what is 25 relevant to sensible substance, such as would provide the same form of substance for all,[105] but he does not allow that, since it is a conglomerate,[106] its proper attributes (*idiômata*) should be such as it itself is. Since here too Aristotle, in wishing to indicate the conglomerate and inauthentic nature of material and composite (*sunthetos*) and generation-bound 30 substance, has given his account of it in terms of change into contraries, so that he views its proprium from the perspective 59,1 of the compositeness and theoretical multiplicity of substance, on the grounds that[107] it ought to go along with its nature and such an entity as this. So if it were the case that it possessed the characteristic of being receptive of contraries in some cases, and in some cases not, then this would have been a mere accident of it; but since acting as a substratum for 5 contraries is true potentially of it in general, and such is the nature of composite substance as to partake of all the forms and to be conglomerated (*sumpephorêsthai*) out of all those which enter into matter, it is only reasonable that receptiveness to contraries should be linked with its composite nature.

[**40**. A problem raised by Plotinus, to the effect that, if we consider substance as being composite, then Matter

[104] We now return, after a considerable gap (from section 14), to the problems raised by Plotinus in *Enn.* 6.1. This one occurs in 6.1.2,15-18, Plotinus' point being that, by stating the proprium (*idion*) of Substance, you have not yet stated its essence (*ti esti*). The *aporia* is reproduced quite closely in Simplicius (115,24-31), though without attribution of authorship (*phasin*, l.24).

[105] Taking *eidos* here to mean 'form' in a non-technical sense, rather than, say, 'species'.

[106] *Sumpephorêmenê*, with somewhat derogatory overtones, borrowed ultimately from Plato (*Phaedr.* 253E 1, *Phileb.* 64E, *Laws* III 693A4). It could be rendered simply 'composition', but that term is needed, I think, to render *sunthetos* below (l.30).

[107] Accepting Spengel's *hôs* for *hôste* here.

turns out to have the best claim to being a substance.]

SELEUCUS: Similarly, Plotinus raises the problem[108] as to how, if sensible substance is never without size or quality, we 10 are to separate off its accidents from it. Let us illustrate our meaning with an example: fire is said to be a substance; if we resolve to remove from it its dryness and its heat, the whole will no longer be substance, but a certain part of it; and this part is its matter, so that only matter will be found to be substance.

DEXIPPUS: First of all, he does not reckon with the mixed 15 and conglomerate nature of substance in this realm, but he separates off and contemplates its various properties on their own, whereas they do not to any degree subsist independently, but are all implicated with each other and have the capacity to constitute a thing at all only in conjunction with each other; and secondly (we would point out) that he separates off heat and dryness and the various differentiae which are constitutive of species as if they were separable 20 qualities, and he does not allow for specific substance, nor for the fact that the primary subsistence (*hupostasis*) of composite substance is from matter and form, as is shown in the *Physics*.[109] For of those things which supervene on substance, the essential qualities complete it and help to constitute it, whereas the others supervene on it in a separable way. It is from these considerations that he proceeded to the notion that matter was substance in the 25 truest sense according to Aristotle.[110]

But that is not so; for he showed in the *Physics* that it is first and foremost form that is substance. Then again, not even the

[108] We move here to *Enn.* 6.3.8, where Plotinus has returned to the analysis of sensible substance, after expounding his own theory of categories of the intelligible world in 6.2. The terminology of 6.3.8 has actually coloured Dexippus' (or Iamblichus') reply to the previous *aporia* (e.g. *hê aisthêtê ousia sumphorêsis tis poiotêtôn kai hulês* 8,19-20).

[109] cf. 1.7, esp. 191a7ff.

[110] Thus far this section is reproduced fairly closely, though not verbatim (and rather more summarily), by Simplicius, 115,32-116,10. After this, however, Simplicius goes on to quote Archytas, with an exegesis (116,11-24), and then to quote Iamblichus himself (116,25-117,2). None of this is represented in Dexippus, nor is the following passage of Dexippus represented in Simplicius, another example of the complex relationship obtaining between these two works. (Of course, the Iamblichean passage which Dexippus omits is a flight of *noera theôria* which Dexippus may well have found unsuitable to this sort of commentary.)

essential qualities which subsist in union with it are separable, but preserve it potentially (for the conjunction of these constitutes the whole entity); but the matter nevertheless remains in a potential state, whether it receives
30 the essential qualities or the separable characteristics. So to regard sensible substance as being inseparable from size and quality is correct so far as it goes, but it does not yet have this absolute truth attached to it, that, even if we separate off[111] the forms in thought, nevertheless Matter remains potentially (characterised by them) since it appears only in conjunction with forms, so that if Matter cannot even exist without the forms, then substance in the truer sense is that constituted by Form.

[**41**. That this is the proprium not only of Substance but also of a proposition and an opinion.]

60,1 SELEUCUS: But is it not so[112] that, even if a proposition (*logos*) and an opinion (*doxa*) do not experience any external influences, they do not therefore not experience (changes) at all? For it undergoes experiences in a way proper for a proposition[113] and receives a change in itself in respect of truth and falsehood; for this is the contrariety relevant to a proposition and to opinion itself.[114] One might as well say,[115]
5 after all, that the soul is not receptive of contraries, since it does not receive white and black (for it is not according to its nature to receive these), but it does receive what it is naturally fitted to receive, e.g. wisdom and foolishness, and for this reason there will be associated with it its own proper contrariety. Even so, then, here too, even as a proposition receives contraries, so does an opinion, that is to say receiving
10 change in respect of truth and falsehood, not, however, change

[111] Reading *khôrizomen* for *khôrizomena* as Busse suggests.

[112] This *aporia* arises out of 4a22ff., where Aristotle responds to just such an objection. It is represented virtually verbatim in Simplicius, 118,26-119,5, but the *lusis* below is quite different. Simplicius refers this *aporia*, and the next one (42), to a mysterious 'he' (*phêsi*, 118,30 and 119,17). Cf. also Porph. 98,7ff. It seems actually to derive from Alexander of Aphrodisias, cf. E. Schmidt, 'Alexander von Aphrodisias in einem altarmenischen Kategorien-Kommentar', *Philologus* 110, 1966, 284ff.

[113] Reading *hôs ho logos* with mss. of Simplicius (118,28), instead of simply *ho logos*.

[114] Or, as Busse suggests, reading *hê autê* for *autês*. 'The same goes for opinion.'

[115] I am not familiar with the idiom *hôra ge legein*, but this, I think, is what it must mean.

in respect of sitting or not sitting (4a24-8). So then there will be associated with them contrariety in respect of change proper to a proposition and an opinion.

DEXIPPUS: To this we will say that it is not in and of itself that a proposition receives truth and falsehood, in so far as it is indicative of these, but it is in the mode of relation, not 15 according to its own nature,[116] that it receives truth and falsehood; for it is by virtue of its being concordant with the facts (*pragmata*) that it is said to be true or false. So then, even as in the case of relatives, with no change taking place in themselves, right becomes left by reason of something else changing its place, so here too it is not by experiencing anything that a proposition, or by changing that an opinion, comes to be false from having been true, but they come to appear different at different times, although remaining 20 unchanged, by reason of the alteration of the facts. But even if one were to grant that such changes are of a proposition *qua* proposition and opinion <*qua* opinion>,[117] the change will be in respect of truth and falsehood, but not in virtue of their experiencing anything; for they remain unchanged.[118]

[**42**. A problem raised by Plotinus[119] to the effect that the stating of its proprium does not set before us what substance is.]

SELEUCUS: Come then, if you will, let us grant that this is the proprium of Substance, still he does not thereby set before 25 us what Substance is.[120] For what touchstone does our mind have for distinguishing substances from attributes (*pathê*)?[121]

[116] Unless one should excise this phrase (as a gloss on *kath' heautên*) as Busse suggests.

[117] Supplying *hôs doxês*, as Busse suggests.

[118] This *lusis*, as I have noted above (n. 106), is not reflected in Simplicius, who has one that is verbally quite distinct, though on the same lines. Only the example of right and left (119,9-10), and the mention of having a relation to *ta pragmata* (119,10-11), concords with Dexippus. Either Simplicius or Dexippus is for some reason choosing to use his own words.

[119] This is not identified as an *aporia* of Plotinus in the body of the text, nor by Simplicius, but in fact reflects Plotinus' objections in *Enn.* 6.1.3, with various verbal reminiscences (e.g. *epereisantes* 60,25 – *epereisasthai* 3,13).

[120] Or, 'the essence (*ti esti*) of Substance'.

[121] This sentence appears virtually verbatim in Simplicius (119,17-18). The rest, though plainly derived from the same source, is not so close (though closer to Plotinus in 18-21), and thus perhaps truer to the Iamblichean original.

If we take the characteristics of being a 'this' and of 'not being in another', this will not embrace the species and the genus; for the species is said of the individual substance, and the genus of both the species and the individual.

61,1 DEXIPPUS: First of all, we will say that it is described as (predicated) of another (*kat' allou*) not in the way that an accident is, because it is dependent on another (distinct thing), but like a genus, which is constitutive of its subject and is said of it in the same way. Then, as we have frequently said 5 before to this man,[122] we add the following, that one must not look for incorporeal substance in the case of bodies, nor that which is in itself and separably one among things that are not able to sustain these characteristics.

One should realise this, that one cannot give accurate definitions of the highest genera,[123] nor is it possible to embrace their essence (*ousia*) completely, for all information about them derives from that which is subsequent to them, 10 and resembles rather a suggestion (*hupomnêsis*) or a rough description (*hupographê*). One should not, therefore, seek more from the account than the degree of clarity of the subject matter allows of, so that it is sufficient to give the proprium, from which one can learn whether any given utterances are predicated of Substance.

[122] i.e. Plotinus, though he has not been mentioned.

[123] This sentence is reproduced verbatim by Simplicius, 119,26-7; the rest of his text (27-30) corresponds more loosely with Dexippus. The remark is actually less than fair to Plotinus, who is quite well aware of this fact, cf. 3.22,18-20.

BOOK 3

[1. That according to Plotinus Quality ought to be ranked before Quantity.]

DEXIPPUS: We have now, Seleucus, best of my companions, dealt in this way with the problems concerning Substance. Do you want us to call a halt at this stage, or shall we transfer our attentions in turn to the nature of Quantity? 64,1

SELEUCUS: Well now, it would not befit any philosopher, let alone you yourself, who can never resist a challenge, to leave this treatise[1] incomplete. Hesiod indeed advises us to bring to perfection the beauty of buildings, lest, as the poet says: 'perching (upon them) a raucous crow may croak';[2] and shall we not fear lest someone should justifiably reproach us with having abandoned our purpose, if we do not carry our treatise to its end, or rather that it should reveal us not to be 'nurselings of Hope' which, Pindar will have it, is 'a sweet accompaniment and fosterage for the heart'.[3] But enough of that. Let us discuss with each other the order of the categories. First of all, then, you must give the reason why Quantity is ranked second after Substance. For Plotinus[4] would have it that Quantity presents manifest indications 5 10 15

[1] A slight lapse from realism here, surely. This is meant to be a dialogue, not a *sungraphê*. This passage, by the way, constitutes evidence for at least Dexippus' intention to comment on the whole *Categories*.

[2] *WD* 747. Dexippus seems to be the only ancient author to quote this line. He likes Hesiod, to judge by his quotation both of this line here and of *WD* 317 at the beginning of the whole work (p. 4,6). This may betoken some degree of Neoplatonic interest in the poem. Proclus later, we should bear in mind, wrote a commentary on it.

[3] Fr. 214 Snell, borrowed by Dexippus from Plato, *Rep.* 1.331A, where it is quoted more fully.

[4] This report does not correspond to anything in *Enn.* 6.1 or 3. In 6.1, in fact, Plotinus seems to accept the order of the categories. It may therefore be evidence of oral teachings of Plotinus. On the other hand, Simplicius, in a parallel passage (122,7ff.), does not make any attribution to Plotinus, and from what follows (122,19ff.) it is apparent that he is taking this from Iamblichus. So Dexippus may have wrongly assumed Plotinus to be the author of an argument left unattributed by Iamblichus. Certainly this objection to the order of the categories is far older than Plotinus.

that it is more akin to Substance, putting the argument in the following terms: we postulate that Substance, in the truest sense, is form, according to Aristotle himself, but if this is so, that which is akin to the form would be prior to those things
20 that are further removed from it. But it is agreed that Quality subsists as a first product of Form, being of the same type as Form itself is, so that if it is linked to Substance in the truest sense, which is that constituted by Form, it would be reasonable that Quality should hold the second rank after Substance.

Most of all,[5] though, because the *quale* (*poion*) is partless
65,1 and non-extended[6] and is distributed about bodies accidentally (*kata sumbebêkos*) without division into parts (*ameristos*), because the *quale* is that which partakes in Quality. It is reasonable, then, in view of this, that Quality should be ranked before Quantity, if indeed the non-extended should be ranked by nature before the extended, the partless before that which has parts, the indivisible, in accordance with the very
5 definition of the *quale*, before the divisible, and the simple before the compound, and is more akin to the incorporeal first principles.

DEXIPPUS: In fact, Archytas the Pythagorean, most assiduous Seleucus, uses this order, in the following words:[7] 'The order of them is as follows: first in order comes
10 Substance, by reason of the fact that this alone serves as a substratum to the rest, and can be conceived of itself by itself, whereas the others cannot be conceived of without it; for they are predicated either of it or (in it) as subject. Second comes Quality; for without there being a "what" there cannot be a "what sort".' Such is the distinction made by the followers of
15 Pythagoras and Plotinus. We however, coming to the defence of Aristotle, give for his ordering some such reason as the following: first of all, along with the really existent and intelligible and one there comes into existence simultaneously Multiplicity and the Forms and models (*paradeigmata*),

[5] This passage is reproduced closely in Simpl. 122,9-11, but in more summary form.

[6] Reading *adiastaton* for *adiaireton*, in accordance with the corresponding passage of Simplicius, as Busse suggests.

[7] This passage, down to end of the quotation of Archytas, is represented in Simplicius more or less verbatim (121,13-18), but what follows that is not. Dexippus here (65,14-66,10) actually seems nearer to Porphyry, cf. 100,11-28, where the argument is produced that Quantity is both more basic to Substance than Quality and has more features in common with it.

which, even if they are united in the intellect of God, nevertheless, though having each a distinct essence and delimitation (*perigraphê*), have their unity comprehended in the simplicity of God, so that, if directly following on Being there arises the number of the Forms, then Quantity[8] takes 20 its start from the primal causes and the noetic realm. Then, if one were to postulate that the multiplicity produced from the One were eliminated (which, indeed, is not extended outward, nor separated from its producer, but has its being in it and about it) – but, if we assumed it eliminated, then the Forms themselves, <to which the Forms>[9] of quality are assimilated, will be eliminated as well. So then, if the distinctive character of Quantity is removed, straightway also the 25 concept of Form is eliminated as well, so that a greater consequence results than those critics intend. All this I have advanced by way of argument, since it was necessary to counter on an intellectual level men of intellect, whose attention is fixed on the intelligible realm, and who are taking account of that realm as well as the sensible; but if we must 66,1 pose Aristotle's intentions more truly, we should rather take our start from this point: the nature of Quantity is more akin to the body and its extension than is that of Quality, so that since there are more features in common between Quantity and Substance than between Quality and Substance, it is reasonable that Quantity should be ranked second after Substance. For instance, the nature of Quantity reveals the 5 divisibility, the compositeness, and the three-dimensionality of Substance, whereas that of Quality reveals only how it is qualified. One has therefore[10] to award the prior degree of affinity to that which has a number of features in common, and not to what has only one, so that since he posited composite and corporeal Substance as primary, it is

[8] Reading *posotês* with Busse (as is necessary), for the *poiotês* of the mss.

[9] Reading *hois ta eidê* before *tês poiotêtos paraphômoiôtai*, as seems necessary for the sense. It will have fallen out by haplography. The latter words are not found in the Laurentianus (A), which is the best manuscript, and Spengel omits them, but some mention of Quality seems required to complete the argument. The idea is that primary Quantity corresponds to the Forms as numbers, and the Forms in their qualitative aspect are logically dependent on these. The argument in fact answers Plotinus on his own terms, e.g. *Enn.* 5.1.5,5ff.

[10] Only here, in this last argument, can some affinity be discerned with the text of Simplicius (cf. 122,25-30), even extending to some verbal echoes.

reasonable for him to rank Quantity with it as being more
10 proper to it and more akin and co-existent with its extension. And we might make this defence, that, since from the point of view of ordinary linguistic usage, the composite is more familiar to us, for this reason the extension that belongs to the composite is more akin to us[11] than is non-extended Quality.

[**2**. A problem raised by Plotinus, in which it is asked what is the element predicated which is common to continuous and discrete magnitude.]

SELEUCUS: Since once again Plotinus[12] raises an objection,
15 saying that one should not assert that both (kinds of Quantity) are alike quantities (for if the continuous is a quantity, then the discrete is not, and if the discrete is, then the continuous will not be), it is incumbent on you to say what it can be that is common to both of them.

DEXIPPUS: To give a succinct answer to the query, they will have in common the element of measure within them, and limit (*peras*), which is to be found both in the discrete and the continuous.[13]

20 But since it is likely that anyone wishing to make difficulties here will declare that quantity in magnitude is quantity by participation in number, and that it takes on the title of quantity only when it is given a number, by being defined, say, as 'two-cubits' or 'three-cubits' long,[14] it is worth while to discuss this question also. The continuous and the discrete in the composition of the cosmos are seen to have
25 each their own peculiar nature; for the discrete is characterised by juxtaposition and 'heaping' (*sôreia*), and the continuous by unity and coherence (*allêloukhia*), and the continuous and unified is called 'magnitude' (*megethos*), while the juxtaposed and discrete is called 'multiplicity' (*plêthos*).

[11] If that is the sense of *proseoikos*. Perhaps 'more akin to Substance'.

[12] At *Enn.* 6.1.4: 'if it is maintained that the continuous is a quantity by the fact of its continuity, then the discrete will not be a quantity. If, on the contrary, the continuous possesses Quantity as an accident, what is there common to both continuous and discrete to make them quantities?' (11,5-8). This passage is loosely paralleled in Simpl. 127,12-29.

[13] This (as Simplicius specifies, 127,14-16) is Plotinus' own answer to the problem, in *Enn.* 6.3.13.

[14] This is a paraphrase of Plotinus' point at 6.1.4,11-14.

On the basis of these distinctions,[15] then, by virtue of the existence of size the cosmos is conceived to be one and is called solid and spherical and connatural with itself, extended[16] and coherent, while by virtue of the form and concept of multiplicity, again, it is conceived to contain the heavenly spheres and the stars and the elements and animals and plants, and in the case of the unified there is division to infinity from the totality, and increase to a limited extent, while in the case of the multiplicity by contrast there is increase *ad infinitum*, and, conversely, division to a limited extent, and both these genera are the objects of their respective sciences, multiplicity of arithmetic and size of geometry; so that if one is to distinguish these from one another in their essence (*huphestêkenai*) and <in their mode of being cognised>[17] then multiplicity would not in itself be Quantity, and magnitude is so accidentally as receiving a specific quantity (*to tosonde*) from number; for what is common to them both is being measured and being limited according to the proper essence (*hupothesis*) of each. 30 67,1 5

[3. That discrete Quantity is prior to continuous, and it abolishes it along with it, without itself being abolished by it, and it is implied by the continuous, but does not imply it.]

SELEUCUS: Well, that is a satisfactory answer to that objection, but I wish to raise another point.[18] It seems to me that the discrete is prior to the continuous, and I would argue

[15] This whole passage, down to 67,3 (reproduced, more loosely, in Simplicius) is to be found virtually verbatim in Iamblichus *De Comm. Math. Sc.*, ch. 7, p. 28,24-29,18 Festa, and at *in Nic* 7,6-22 Pistelli. Iamblichus obviously also used the same material in his *Categories* commentary, but which came first cannot be determined. Iamblichus has a propensity to quote himself repeatedly. Here Dexippus is being somewhat more faithful to the original than Simplicius.

[16] Reading *diatetamenos* for mss. *diatetagmenos*, following Iamblichus (and Simplicius).

[17] Adding *noeisthai* here, with Busse.

[18] This argument from priority and posteriority may arise from that of Plotinus in *Enn.* 6.3.13,12-15, as Busse suggests, but it is distinct from it. Plotinus makes the point that Quantity is essentially Number and numbers, according to Aristotle himself in *Metaph.* 3.999a6-14, can have no common genus, since each smaller number is prior to the next one. Dexippus transposes this argument to the priority of the discrete to the continuous. This whole *aporia* is reproduced in Simpl. 126,6-10 and 126,23-127,11.

10 that as follows: the continuous is divisible into parts which are always further divisible, so that if the divisible did not exist, neither would it; so divisible quantity is prior to continuous. But if it is prior, then Quantity itself would not be common to both; for in the case of things which have a relation of prior to posterior there can be no common genus; for species do not abolish (*sunanairei*) each other, and no abolition of a 15 species is sufficient to abolish the genus, but in cases where priority and posteriority can be observed, the one is shown to be destructive of the other.

DEXIPPUS: But if we are prepared to look at the position honestly, in one sense we will grant them priority and posteriority, but in so far as they partake of the same genus we will discover that they enjoy an equal degree of community. For nothing prevents one and the same account, 20 viewed as completing each of the species in relation to the genus, from being simultaneously present to all of them, as many as are capable of participating equally in it through themselves,[19] so that even if some are prior and others posterior, and they possess in themselves the nature of sharing immediately in the genus, the genus will be present to both, providing the same participation in itself to those things 25 able to avail of such participation. In the same way, after all, although assertion is by nature prior to negation, when we focus on them in the process of expressing opinions (*apophainesthai*) and true and false statements, we conceive of them as falling under the same genus, recognising that in one way they possess priority and posteriority, but also that, in so far as they share in the same genus, they admit of being 30 divided off as coordinate species. But that Aristotle held such a view about this matter, on the one hand taking number to be prior in order, but on the other hand making an equal division between the discrete and the continuous, gathering both in the one genus, he has made plain from the order of his 68,1 presentation. For he lists the discrete before the continuous both in the first division he makes and in the enumeration of the particular types (of quantity), giving an account of those falling under magnitude. For because we seem to comprehend

[19] Reading *hautôn* for *autôn* as Busse suggests (cf. *di' heauton* in the next line).

quantity in the mode of magnitude through number, for this reason number is placed first, even as, because we come to recognise negation through assertion, for that reason it too is 5 placed first, while when one is making an equal division neither is deprived of community of genus. And so here too, it is with regard to their equal contribution to the definition of the genus, and their equal suitability for this, that he has set them over against each other. Furthermore, the process of being divided is distinct from the state of being divided so that, if the potential is comprehended in the actual (for the 10 continuous is potentially divisible), we would not be correct in assuming the divisible to be already existent; for what is potentially in a state is all the more so posterior to what is actually in that state.

[4. A problem raised by Plotinus, to the effect that the nature of Number is different in itself from what it is when it is related to specific quantity.]

SELEUCUS: Well then, let that be our answer on the question of the division of Quantity in general. I want now, though, to add some particular difficulties which Plotinus raises about each of the species falling under the genus of Quantity. First of all, he makes a distinction about Number[20] 15 and declares that its essence is different in itself from what it is when it is associated with specific quantity. For it has one essence when considered on its own, which is ignored by Aristotle (as Plotinus maintains), while it is the other that is manifested when we call a Number a quantity.

DEXIPPUS: If we subject the categories to the test of Plotinus' structurings (*diataxeis*) and Plato's assumptions (*hupotheseis*), then the argument as stated would have some 20 plausibility, but if we are prepared to follow Aristotle, we will certainly not divide off number when considered as a

[20] This is an interpretation, at least, of Plotinus' point about Number in *Enn.* 6.1.4,23ff., where he declares that Number in itself (that is, Ideal Number) is a substance, while numbers in objects are in a paradoxical situation; as numbers they are still, surely, substances, though as enumerated objects they may be regarded as quantities. The whole passage is paralleled loosely in Simpl. 129,8-27, but closely enough to indicate that the same source is being followed. It is not clear, however, which of the two is sticking closer to the original.

quantity. For Aristotle does not hold that number possesses its essence as one distinct thing,[21] and its being a quantity as another, but he says that this is a proprium of composite and material things and such things as have their form
25 particularised (*en merei*); but such things as are immaterial and non-composite forms he shows in all his treatises (*akroaseis*) to be identical in their essential being and in their existence and manifestation in what receives them. So if Plotinus is distinguishing intelligible numbers from those numbers that participate in them, and defining separately the particular characteristics of each, he is simply laying out his
30 own assumptions; but we will remind ourselves that according to Aristotle numbers belong to the very essence of things, and are neither prior nor posterior to them, but are
69,1 entirely coordinate with them. So according to him they are in no way absent from anything that is in any way able to participate in them, but are present to all in a way proper to their own indivisible essence, while exhibiting variation according to the various natures of the receptacles, so that being both numbers and specific quantities coexists in the things numbered.

[**5**. A problem raised by Plotinus, as to whether number subsists in numbered things, or measures them from outside.]

5 SELEUCUS: Following on from the same problem, Plotinus also raises the question, does Number subsist in numbered things, or does it measure them from outside, like a ruler?[22] For if it subsists by itself, then it will be a measure, but no longer a quantity; whereas if it subsists in (the things numbered), it will itself be measured by something else.

[21] On the basis of the parallel Simplicius passage (129,14-15), Busse thinks that the distinction here should be between number as an indefinite quantity (reading <*poson*> *einai*) and number as a *definite* quantity (*tosonde*). This distinction is indeed made in Simplicius, but it does not seem necessary to the sense here. It may be a subtlety introduced by Simplicius.

[22] This refers to *Enn.* 6.1.4,25ff., which does indeed follow on directly from his previous *aporia*. Simplicius reports the *aporia* at 130,8-19, first paraphrasing Plotinus (9-14), and giving what he explicitly presents as Iamblichus' reply to him (15-19). Since this agrees virtually verbatim with what we have in Dexippus (9-18), we get a particularly clear idea of what is going on.

DEXIPPUS: But it is worth saying again[23] what I said at the beginning of this: we declare that Number is present to the things numbered and subsists along with them, but yet it does not have its existence (*hupostasis*) in them in the role of an accident. For just as the forms-in-matter have on the one hand each their own substance and existence, but yet coexist with matter, so too numbers, while existing in their own substances, are conjoined with the things which they delimit and upon which they impose their proper measure, coexisting with the things numbered as numbering agents, for even as the forming agents (*eidopoiounta*) are ranged with the things informed, primarily as being *in themselves*, but they produce extension (*diastasis*) in virtue of their substratum, so too do the numbering agents, when <ranged with>[24] the things numbered, relate each of them to their proper measure while remaining true to their own essence, and receiving no effects from the things measured. Let us make our position still clearer by an example;[25] the form of the statue has a reason-principle proper to it on its own, but it is viewed in conjunction with its matter. And so number is united with things numbered while possessing its own proper existence (*hupostasis*).

[**6**. That Quantity, as such, is not by any means a quantified thing (*poson*).]

SELEUCUS: But the quantified thing, he says, would not be Quantity.[26]

DEXIPPUS: Well, if he had anything else to show that

[23] This should presumably refer back to 68,19ff., but I can see nothing there that it can refer to.

[24] Adding *suntattomena* to govern the dative here, as Busse suggests (cf. *suntattetai* in l. 18 above), though it *could* perhaps be understood.

[25] This final passage sounds like an addition by Dexippus. It is not represented in Simplicius.

[26] This is interesting, since Plotinus himself does not say this, at 6.1.4,32ff., but only 'As for numbers themselves, why are they Quantity (*poson*)?' Following on this sentence, however, in his paraphrase, Simplicius (reporting Iamblichus) adds the sentence, 'for Quantity itself is not by any means a quantified thing', which is echoed, strangely enough, most closely not by Seleucus' statement, but by the prefatory heading, which would seem to confer some authority on these headings. Either, then, something has fallen out of our text of Plotinus, or Iamblichus has added an explanatory gloss, and Dexippus and Simplicius are following him rather than reading Plotinus for themselves. Simplicius continues to quote Iamblichus here

Quantity was besides a qualified thing, he would have been justified in stating this. But since we have been contending from the outset that nothing else is the dominant element of Quantity but measure, and this is common to Quantity and
30 the quantified, he cannot reasonably refuse to say that Quantity is quantified. One might come to the same conclusion from the fact that no other predication can be attached to (*hupotheinai*) the nature of the quantified, but it is seen as an independent principle, connected with none of the other genera, so that Number too, whether in itself or in its participants, whether co-subsisting with them or actually
35 preserving its own nature while co-subsisting, whether subsisting in the role of counting agent or of counted thing, would in all cases be of the essence of the quantified.

[7. A problem raised by Plotinus, to the effect that language (*logos*) is an accident of Quantity.]

SELEUCUS: So then, we have thus solved the problems raised by Plotinus about Number. Let us turn next to language, then, and see what problems he raises in that
70,1 connexion.[27] He says that language possesses a particular quantity while having its own peculiar mode of existence[28] and is quantified to the extent that it has quantity as an accident (*sumbebêkos*); whereas in so far as language is meaningful (*sêmantikos*) it has as its material air, having acquired its subsistence through its impact (*plêgê*) on this; so whether language in utterance (*phônê*) is an impact, or not a simple impact, but an imprinting (*tupôsis*) of the air, as if
5 shaping it, in either case it would be a significatory 'making' (*poiêsis*), and so the language in utterance will belong to the category either of 'making' or of 'being affected', or to both

(130,19-24), while Dexippus turns to a loose paraphrase. The *lusis* of the *aporia*, we may note, once again answers Plotinus from himself (cf. 6.1.4,35ff.).

[27] This refers to *Cat.* 4b32-7, where *logos* is stated to be one of the types of discrete Quantity. Plotinus makes his criticisms in *Enn.* 6.1.5,2-14. This passage is discussed at length by Henry, op. cit., 248ff. I see no necessity, however, to postulate an oral source over and above *Enn.* 6.5, despite the variations in phraseology which he lists.

[28] This expands on Plotinus' very laconic *alla logos ôn tososde esti*, which presumably means 'but as being speech it is of a certain quantity', which is extremely obscurely expressed. Voss actually proposed to insert *phônê* before *ôn*, which would give the meaning 'speech, *qua* utterance, is of a certain quantity'.

'making' and 'being affected', but not to Quantity. What are we to say to that?

DEXIPPUS:[29] That utterance is not simply an impact on air (for a finger, when waved, makes an impact on air, but does 10 not yet produce a sound), but a certain quantity and intensity of impact, and indeed of such a quantity as to be audible, and that which is equated to the measure of the hearing faculty and has in itself excess and deficiency. So it is in virtue of the magnitude of the impact and through postulating quantities as the measure of the voice (*phônê*) that he declares, reasonably, that it is a quantity.

[8. A problem raised by Plotinus, to the effect that it is not through itself, but as being said to be in time, which is a quantity, that language is considered itself to be a quantity.]

SELEUCUS: But since Plotinus says[30] that language is said 15 to be in time, which is a quantity, it itself too is considered to be a quantity; so then Quantity will be in time and through time.

DEXIPPUS: But we say to this[31] that there is a smallness and magnitude of the voice according to its nature, or through the smallness or magnitude of the windpipe; for when the individual sound (*phônêma*) is divided up and broadened out, 20 it produces an extended sound (*plêthos*), while in so far as it is constricted, a small one (*oligotês*). That which is naturally

[29] This whole reply of Dexippus is given by Simplicius, 131,10-16, as a verbatim quotation from Iamblichus, showing excellently the relationship between the two works. Dexippus has modified Iamblichus slightly, e.g. simplifying the rather Iamblichean double compound *sunexizoumenê*, 130,23-4, to *exizoumenê*.

[30] There is a mystery here. What seems to be referred to is Plotinus' next *aporia* in 6.1.5 (ll. 14-26), which is no longer about *logos*, but about time itself – not time as a measure of language. Either Iamblichus (and Porphyry before him) has misunderstood the text of Plotinus, or we have here evidence of another text than ours. (An alternative, as Steven Strange suggests to me, is that the phrase in 1.2, *metreitai men gar*, 'for it is certainly measured', is being taken as an indication that language is in time.)

[31] A version of this passage is presented by Simplicius, 131,27-132,6, as a view, not of Iamblichus, but of Porphyry. Rather than conclude, however, that Dexippus is following Porphyry (whatever about Simplicius), I would prefer to postulate that Dexippus is following Iamblichus, who is adapting Porphyry. The treatment of *logos* in Porphyry's short commentary, 101,24-102,9, is actually rather different, which adds to our problems.

short is said in a correspondingly short time, while that which is long in a long time, as if language had the quantified aspect of speech in itself, prior to the quantified aspect of time; for even if I am not uttering language, still the written language
25 possesses by nature long and short syllables. And we also know this, after all, that it is possible both to utter a short syllable over a long time-interval, and a long one in a short interval. For anyone skilled in metres will introduce alterations for both metrical and rhythmical reasons, and actually changes individual letters, as for instance *xeros* for *xêros* (dry) and *Diônysos* for *Dionysos*.[32] But if it is obvious that there is a distinction between natural length and
30 temporal length, then it is not by reason of covering a short space of time that a syllable is said to be short, but it is by reason of the fact that it is short that it is said to cover a short space of time.

[**9**. A problem raised by Plotinus, to the effect that it is not sufficient to be measured to be accounted a quantity.]

SELEUCUS: But again, what would we say to the man's[33] objection, which does not accept that being measurable is sufficient to be accounted a quantity. For after all, he says, a piece of wood is measured by the cubit, but one would not describe it as a quantity, simply because it is measured, but
35 because it is a magnitude, and it is in virtue of the quantity that is accidental to the wood that it is measured by the
71,1 quantity which is accidental to the cubit-long wood, but wood as such is not measured by wood. And so in the case of language also, it is not measured *qua* language, but in so far as it is in time, so that time would be what measures time, the

[32] Both these examples are taken from Homer: *xeros* for *xêros*, in *Od.* 5, 402; *Diônysos* in *Iliad* 6, 132, etc. They come from Porphyry (cf. Simpl. 132,4).

[33] sc. Plotinus, as the index of contents specifies, but again, there is a mystery about this, since neither in 6.1.5 or in 6.3.11 is this point made, unless it is being derived from the final passage of 1,5 about the equal and the unequal. Simplicius, in discussing *Cat.* 6a26-35, where the property of being equal and unequal is discussed, uses a piece of wood as an example (151,22-3), but he does not do this in reference to an *aporia* of Plotinus (though Plotinus does use wood as an example in 6.3.11,5). Nevertheless, it seems possible that this discussion (in Porphyry or Iamblichus) arose from this passage, even though it is not picked up as such by Simplicius. It could be an oral *aporia* of Plotinus – or an original contribution by Dexippus.

lesser unit measuring the greater.

DEXIPPUS: To this we will make use of the same responses, for if speech contains within itself an innate magnitude, according to which musicians are accustomed to mark intervals in it, it will be a quantity in itself, and not accidentally. For everything either measurable or measuring is a natural quantity, and so this, as being measured by time, would be a quantity.[34] 5

[**10**. Another problem raised by Plotinus, to the effect that even as action is only accidentally a quantity, so too is language.]

SELEUCUS: Again the same man raises another problem,[35] to the effect that action (*praxis*) is said to be great or small by reference to the time it takes, and so too language, as being measured by intervals, would be only accidentally quantified. 10

DEXIPPUS: In reply to this it must be said that, of letters (*stoikheia*) some are naturally long, others naturally short. So then there will naturally inhere in speech (*phônê*) both magnitude and brevity. 15

[**11**. That language, being a thing involving combination (*sumplokê*) cannot fall under the categories, at least if the categories are simple items (cf. *Cat.* 1a16, 1b25).][36]

[**12**. That he should rather have described language and number as continuous, than body.]

[**13**. Why does he seem doubtful when he says 'if five is part of ten' (4b26-27).][37]

[34] The ms. A breaks off here; mss. CMR gives the final fragmentary passage.

[35] Again, this does not relate closely to anything in *Enn.* 6.1.5. It may indeed be an oral *aporia* transmitted by Porphyry.

[36] This section is discussed extensively by Henry, op. cit., 242ff. It certainly corresponds to nothing in *Enn.* 6.1, and so may be an oral *aporia*. It is mentioned briefly by Simplicius 130,32-3.

[37] cf. Simplicius, 133,6-10. The answer is that he is not in doubt about this, but only recognising that there are other possible ways of dividing ten. This, for Simplicius, is the end of the discussion of discrete quantity (though Dexippus, section 12, seems to concern the continuous).

[14. That body and magnitude and line are not quantities in themselves, but only by participation in quantity.][38]

[15. That although he has not previously explained place and time, he discusses them as if they were already familiar.][39]

[16. That he has left out another, third type of Quantity.][40]

[17. A difficulty raised by Plotinus about Time, as to how it is said to be a Quantity.][41]

[18. That time belongs rather to the category of Relation.][42]

[19. A problem raised by Plotinus about Space.][43]

[20. Whether two or three characteristics belong to things which have position.][44]

[21. Whether Place and Plane Surface have position in

[38] cf. Simplicius 135,11-34, but not dealt with in exactly these terms.

[39] cf. Simplicius 135,25-134,11. The answer is that he is not claiming here to present their essence. That is a physical question, and he deals with it in his *Physics*. In a logical context, we can be content with *koinê ennoia*.

[40] This may refer to the question raised by Lucius and Nicostratus and replied to by Iamblichus, whether a third type of Quantity, weight (*rhopê*), should not be added to the two, discrete and continuous, given by Aristotle, and replied to by Iamblichus, who actually is in favour of adding *rhopê*, on the authority of Archytas. All this is reported by Simplicius back at 128,5-129,7, and not dealt with by Dexippus earlier.

[41] This *aporia* is raised in *Enn.* 6.1.5,14-20, and returned to in 6.3.11. Simplicius discusses the question at 134,12-24.

[42] cf. Simplicius 134,25-32. If someone wishes, as Plotinus seems to at 6.3.11,7ff., to refer Time, as the measure of motion, to the category of Relation, one must reply to him that 'measure' may be understood in two senses, absolute and relative, and Time is a measure in the former sense.

[43] Mentioned by Plotinus, not in 6.1.5, but in 6.3.11,6ff., where he describes it as 'circumscribing Body' (*periektikos sômatos*). Simplicius discusses this at 134,33-135,7.

[44] cf. Simplicius, 136,12-137,29. The case for three characteristics (i.e. place, co-existence of parts, and continuity of parts) was first propounded, it seems, by Porphyry (cf. 104,12ff.), Iamblichus (ap. Simpl. 136,20ff.), disputes that *place* is necessary, since the line does not have place but has position, and this is doubtless the position we would find in Dexippus.

the primary sense (*proêgoumenôs*).][45]

[22. That in general there can be no such thing as position.][46]

[23. That plane surface is only accidentally a quantity.][47]

[24. That some things that have position are not quantities.][48]

[25. What is the criterion of definite and of indefinite quantity?][49]

[26. Why is indefinite quantity said to be sometimes absolute and sometimes relative?][50]

[27. That there is opposition in the category of Quantity, and this even though Aristotle says that 'a quantity has no contrary' (5b11).][51]

[45] This question does not seem to be discussed as such by Simplicius, but both Porphyry (105,12ff.) and Iamblichus, following him (ap. Simpl. 139,24ff.), make a distinction between *posa proêgoumenôs*, which may lie behind this, and Simplicius himself, at 141,3-9, contrasts Time and plane surface as *proêgoumenôs posa* with certain 'indefinite predicates' which are *kata sumbebêkos*.

[46] cf. Simplicius 140,22-31, described as an *aporia* of 'the same men' (perhaps referring to Lucius and Nicostratus). The *aporia* really is that nothing can have position, if that only can have position the parts of which remain constant, and the parts of everything are in continued flux. The answer is that there *is* something which remains constant, whether we call it the 'secondary substratum' or the 'particular quality' (the Stoic *idiôs poion*).

[47] cf. Simplicius 140,31-141,3.

[48] Not dealt with as such in Simplicius.

[49] cf. Simplicius 144,27-31: 'Definite quantity is that which is circumscribed and conceived of according to a definite measure, as for instance "two" or "three" or "one cubit long"; indefinite is what is without circumscription and left uncertain according to what standard of excess it exceeds, or of deficiency that it falls short; examples are "many" or "few" and "the great and small".' Cf. Porph. 108,13ff.

[50] This follows from the previous passage in Simpl. 144,31-145,9. The answer is that it is sometimes referred to something else of the same type (*homogenes*), and sometimes not. Simpl. then quotes a 'transcendental' explanation of this by Iamblichus (145,10-146,21), but one cannot be sure that Dexippus adopted this.

[51] cf. Simplicius 147,25-148,36. This is actually an *aporia* arising out of *Cat.* 6a11-18, which appears to be in contradiction to 5b11, but as Simplicius (doubtless following Iamblichus and Porphyry) explains, is not. For Porphyry, cf. 107,1-30, where he includes a reply of Herminus to this problem, not mentioned as such by Simplicius. For the problem, cf. the beginning of *Enn.* 6.3.12.

[**28**. That 'much' and 'little' can be quantities.][52]

[**29**. Should shapes be called qualities or quantities?][53]

[**30**. Where should geometry be ranked, within Quantity or Quality?][54]

[**31**. How can he say that there is no contrary to a quantity?][55]

[**32**. That place is not in itself the cause of quantitative movement.][56]

[**33**. Why do we not call the diameters (of a circle) contraries, if we measure contrariety by the greatest degree of separation (*diastasis*)?][57]

[**34**. Are we to speak of equal and unequal in relation to heaviness and lightness or not?][58]

[**35**. That if the equal and unequal are said of the heavy and the light as of qualities then this would not be a

[52] This is in reference to Aristotle's claim at 5b14ff. that 'much' and 'little', 'great' and 'small' are not really expressions of quantity but relativity. Iamblichus has a good deal to say, however, in favour of such terms denoting, not relatives, but indefinite quantities, ap. Simpl. 144,7-146,21, some of which we must presume to have appeared in Dexippus; cf. also Porph. 107,31-108,17, arguing in defence of Aristotle's position.

[53] No trace of this or the following question being discussed by Simplicius under Quantity, but we can see from the latter in the Commentary that he ranked both shape (cf. 227,28) and geometry (226,5) under Quality, so this is probably where Dexippus assigned them.

[54] This may relate to Plotinus' discussion in *Enn.* 6.3.14. Nothing in Simplicius.

[55] We seem here to be backtracking somewhat, to question 27; cf. above, n. 49.

[56] This is dealt with briefly at Simpl. 149,22-6. The issue is that, if it was, then one could say that place in itself constituted a defining bound for things in motion. But in fact it is the other way around.

[57] cf. Simpl. 149,1-10. The reason is that diameters, though spatially at a maximum distance, are on the cosmic level not opposed, whereas the centre is specifically opposed to the circumference.

[58] This seems to arise (to judge from Simplicius' presentation of it, 151,32-152,31) from the assertion at 6a26ff., that equal and unequal can only be predicated of heavy and light 'improperly' (*katakhrêstikôs*). This is refuted by 'the more recent commentators' (152,13), who adduce the authority of Archytas, thus pointing to their identity with Iamblichus (though the distinction is made already by Porphyry, 110,32).

proprium of quantity.][59]

[**36**. Why did he say 'but of the rest, whatever is not a quantity would certainly not seem to be called equal and unequal' (6a30-2)?][60]

[**37**. That even if equal and unequal are said of Quantity, this would not be a proprium of Quantity.][61]

[**38**. How comes it that, having removed contrariety from Quantity (5b11), he yet postulates contrariety in its proprium (6a11)?][62]

[**39**. That it is not true of every quantity that equal and unequal are said of it.][63]

[**40**. Where should one place the negations of Quantity?][64]

59 This follows from the previous question, and is discussed in Simpl. 152,23-32.

60 This is covered in Simpl. 151,23-31, where it is explained that 'equal' and 'unequal' in the case of such things as 'white' or 'state' (*diathesis*) are used not *katakhrêstikôs* (cf. n. 58 above).

61 This does not seem to be dealt with as such by Simplicius, but for this and the following *aporiai*, cf. *Enn.* 6.1.23ff.

62 cf. Simpl. 153,8-18. This *aporia* refers to the inserting of the contraries 'equal' and 'unequal' as part of the *proprium* of Quantity. The answer is that they are not being regarded here as contraries, but simply as complementary relatives.

63 cf. Simpl. 153,19-154,2. The problem concerns such entities as the unit or point, which have no extension, and thus cannot be said to be equal or unequal, if the definition of these involves having an equal or unequal number of *measures*. A possible answer, supplied by Simplicius, is that at least unit and points can be equal to *each other* (though not, presumably, unequal).

64 cf. Simpl. 155,4-14. The answer is that negations (as is the case with derived forms, such as adverbs) belong in the category of their positives. This (apart from a general comment, quoted from Iamblichus 155,15-28, on how to identify the propria of categories) ends Simplicius' discussion of Quantity. He has been following the source of Dexippus pretty closely.

Textual Emendations

The following textual changes from Busse's edition have been adopted in the translation (in many cases following conjectures of Busse himself in his *apparatus criticus*).

5,25 — <*kai tôn pragmatôn*>, after *lexeôn*
8,11 — <*ei*> after *all'*
12,7 — *xustropoiei* (conj. Busse), for *xustropôn* of mss.
12,26 — *proeirêmenôi* for *proeirêmenôn*, excising *tôn nun* before it (conj. Busse).
15,7 — Excise *posa* after *esti* (conj. Busse).
15,8 — Insert <*hoion*> before *diplasion*, and after it, *exempli gratia* <*kai triplasion· tauta gar esti hôs pollaplasia·*>
16,1 — *exêrtêtai*, for *exêirêtai*, with ms. A.
20,15 — *idiôn* for *idian*.
26,2 — Read *hôs kath' hupokeimenou legomena kat' ousian proslambanontes* for *hôs kat' kath hu. l. k. ou. p.* (misprint in Busse's text).
26,4 — Read *kata ton auton logon* for *kata ton hautou* (*autou* mss.) *logon*.
26,29-31 — Comma after *rhêthêsontai*; full stop after *katêgoroumenê*.
28,4 — *tois hupokeimenois* for *tois autois* (conj. Busse).
28,13 — <*ho*> before *kai* (understanding *tattetai* after *skhêma*).
33,14 — *anoplein* for *anapnein*, with Felicianus (conj. Busse).
34,1 — *elegomen* for *legomen*.
41,6 — *prolabôn*, with mss. A and M, for *proslabôn*.
45,17 — *pantôn* for *pantôs*.
51,24 — *hoti men* for *hoti hê en*.
52,27 — Insert <*ei*> before *to eidos*.
52,28 — Excise [*tôi eidei*], cf. Simpl.
53,10 — *mê parallattêi ti* for *hêi parallattei*, cf. Simpl. (conj. Busse).
53,13 — Insert <*diaphoran*> after *heuriskontes* (conj. Busse).
54,8 — Insert <*kata gar tên anthrôpotêta*> after *anthrôpotêta*, with Felicianus (conj. Busse).
57,5 — Insert <*an*> after *êdunato*, with Spengel.
58,7 — *exarkei* for *exêrkei* (cf. Simpl.).
59,1 — *hôs* for *hôste*, with Spengel.
59,33 — *khôrizomen* for *khôrizomena* (conj. Busse).
60,2 — Insert <*hôs*> before *ho logos* (cf. Simpl.).
60,21 — Insert <*hôs doxês*> after *doxês* (conj. Busse).

65,1	*adiastaton* for *adiaireton*, cf. Simpl. (conj. Busse).
65,24	Insert <*hois ta eidê*> before *tês poiotêtos*.
66,30	*diatetamenos* for *diatetagmenos*, cf. Iambl. ap. Simpl. (conj. Busse).
67,22	*hautôn* for *autôn* (conj. Busse).
69,20	Insert <*suntattomena*> after *arithmêtois* (conj. Busse).

Bibliography

Anton, J.P., 'Ancient interpretations of Aristotle's doctrine of homonyma', *Journal of the History of Philosophy* 7, 1969, 1-18.

Aubenque, P., 'Plotin et Dexippe, exégètes des Catégories d'Aristote', in C. Rutten & A. Motte (eds), *Aristotelica. Mélanges offerts à Marcel de Corte*, Brussels-Liège 1985, 7-40.

Boeft, J. den, 'Desippo', *Enciclopedia di Filosofia*, vol. 2, cols 391-2.

Busse, A., 'Der Historiker und der Philosoph Dexippus', *Hermes* 23, 1888, 402-9.

Hadot, P., 'The harmony of Plotinus and Aristotle according to Porphyry', in R. Sorabji (ed.), *Aristotle Transformed*, London and Ithaca N.Y. 1990, 125-40.

Henry, P., 'Trois apories orales de Plotin sur les Catégories', *Zetesis, Mélanges de Strycker*, 1973, 234-65.

Henry, P., 'The oral teaching of Plotinus', *Dionysius* 6, 1982, 4–12.

Hoffman, P., 'Catégories et langage selon Simplicius – la question du "*skopos*" du traité aristotélicien des *Catégories*', in *Simplicius: sa vie, son oeuvre, sa survie. Actes du colloque international de Paris, 1985*, Berlin 1987.

Kroll, W., 'Dexippos', *RE* 5, 1905, cols 293-4.

Moraux, P., *Der Aristotelismus bei den Griechen*, Berlin, vol. 1, 1973; vol. 2, 1984.

Noica, C., *Porfir, Dexip, Ammonius. Comentarii la Categoriile lui Aristotel, însotite de Textul Comentat Traducere, Curînt înainte si Note*, Bucarest 1968.

O'Meara, D., *Pythagoras Revived: Mathematics and Philosophy in late Antiquity*, Oxford 1989.

Schmidt, E., 'Alexander von Aphrodisias in einem altarmenischen Kategorien-kommentar', *Philologus* 110, 1966, 277-86.

Tweedale, M., 'Alexander of Aphrodisias' views on universals', *Phronesis* 29, 1984, 279-303.

Usener, H., Review of L. Spengel (edition of Dexippus), *Litterar. Centralblatt*, 1860, 124-5.

Appendix
The Commentators*

The 15,000 pages of the Ancient Greek Commentaries on Aristotle are the largest corpus of Ancient Greek philosophy that has not been translated into English or other modern European languages. The standard edition (*Commentaria in Aristotelem Graeca*, or *CAG*) was produced by Hermann Diels as general editor under the auspices of the Prussian Academy in Berlin. Arrangements have now been made to translate at least a large proportion of this corpus, along with some other Greek and Latin commentaries not included in the Berlin edition, and some closely related non-commentary works by the commentators.

The works are not just commentaries on Aristotle, although they are invaluable in that capacity too. One of the ways of doing philosophy between A.D. 200 and 600, when the most important items were produced, was by writing commentaries. The works therefore represent the thought of the Peripatetic and Neoplatonist schools, as well as expounding Aristotle. Furthermore, they embed fragments from all periods of Ancient Greek philosophical thought: this is how many of the Presocratic fragments were assembled, for example. Thus they provide a panorama of every period of Ancient Greek philosophy.

The philosophy of the period from A.D. 200 to 600 has not yet been intensively explored by philosophers in English-speaking countries, yet it is full of interest for physics, metaphysics, logic, psychology, ethics and religion. The contrast with the study of the Presocratics is striking. Initially the incomplete Presocratic fragments might well have seemed less promising, but their interest is now widely known, thanks to the philological and philosophical effort that has been concentrated upon them. The incomparably vaster corpus which preserved so many of those fragments offers at least as much interest, but is still relatively little known.

The commentaries represent a missing link in the history of philosophy: the Latin-speaking Middle Ages obtained their

* Reprinted from the Editor's General Introduction to the series in Christian Wildberg, *Philoponus Against Aristotle on the Eternity of the World*, London and Ithaca N.Y., 1987.

knowledge of Aristotle at least partly through the medium of the commentaries. Without an appreciation of this, mediaeval interpretations of Aristotle will not be understood. Again, the ancient commentaries are the unsuspected source of ideas which have been thought, wrongly, to originate in the later mediaeval period. It has been supposed, for example, that Bonaventure in the thirteenth century invented the ingenious arguments based on the concept of infinity which attempt to prove the Christian view that the universe had a beginning. In fact, Bonaventure is merely repeating arguments devised by the commentator Philoponus 700 years earlier and preserved in the meantime by the Arabs. Bonaventure even uses Philoponus' original examples. Again, the introduction of impetus theory into dynamics, which has been called a scientific revolution, has been held to be an independent invention of the Latin West, even if it was earlier discovered by the Arabs or their predecessors. But recent work has traced a plausible route by which it could have passed from Philoponus, via the Arabs, to the West.

The new availability of the commentaries in the sixteenth century, thanks to printing and to fresh Latin translations, helped to fuel the Renaissance break from Aristotelian science. For the commentators record not only Aristotle's theories, but also rival ones, while Philoponus as a Christian devises rival theories of his own and accordingly is mentioned in Galileo's early works more frequently than Plato.[1]

It is not only for their philosophy that the works are of interest. Historians will find information about the history of schools, their methods of teaching and writing and the practices of an oral tradition.[2] Linguists will find the indexes and translations an aid for studying the development of word meanings, almost wholly

[1] See Fritz Zimmermann, 'Philoponus' impetus theory in the Arabic tradition'; Charles Schmitt, 'Philoponus' commentary on Aristotle's *Physics* in the sixteenth century', and Richard Sorabji, 'John Philoponus', in Richard Sorabji (ed.), *Philoponus and the Rejection of Aristotelian Science* (London and Ithaca, N.Y. 1987).

[2] See e.g. Karl Praechter, 'Die griechischen Aristoteleskommentare', *Byzantinische Zeitschrift* 18 (1909), 516-38; M. Plezia, *de Commentariis Isagogicis* (Cracow 1947); M. Richard, *'Apo Phônês', Byzantion* 20 (1950), 191-222; É. Evrard, *L'Ecole d'Olympiodore et la composition du commentaire à la physique de Jean Philopon*, Diss. (Liège 1957); L.G. Westerink, *Anonymous Prolegomena to Platonic Philosophy* (Amsterdam 1962) (new revised edition, translated into French, Collection Budé, forthcoming); A.-J. Festugière, 'Modes de composition des commentaires de Proclus', *Museum Helveticum* 20 (1963), 77-100, repr. in his *Études* (1971), 551-74; P. Hadot, 'Les divisions des parties de la philosophie dans l'antiquité', *Museum Helveticum* 36 (1979), 201-23; I. Hadot, 'La division néoplatonicienne des écrits d'Aristote', in J. Wiesner (ed.), *Aristoteles Werk und Wirkung* (Paul Moraux gewidmet), vol. 2 (Berlin 1986); I. Hadot, 'Les introductions aux commentaires exégétiques chez les auteurs néoplatoniciens et les auteurs chrétiens', in M. Tardieu (ed.), *Les règles de l'interprétation* (Paris 1987), 99-119. These topics will be treated, and a bibliography supplied, in a collection of articles on the commentators in general.

uncharted in Liddell and Scott's *Lexicon*, and for checking shifts in grammatical usage.

Given the wide range of interests to which the volumes will appeal, the aim is to produce readable translations, and to avoid so far as possible presupposing any knowledge of Greek. Footnotes will explain points of meaning, give cross-references to other works, and suggest alternative interpretations of the text where the translator does not have a clear preference. The introduction to each volume will include an explanation why the work was chosen for translation: none will be chosen simply because it is there. Two of the Greek texts are currently being re-edited – those of Simplicius *in Physica* and *in de Caelo* – and new readings will be exploited by translators as they become available. Each volume will also contain a list of proposed emendations to the standard text. Indexes will be of more uniform extent as between volumes than is the case with the Berlin edition, and there will be three of them: an English-Greek glossary, a Greek-English index, and a subject index.

The commentaries fall into three main groups. The first group is by authors in the Aristotelian tradition up to the fourth century A.D. This includes the earliest extant commentary, that by Aspasius in the first half of the second century A.D. on the *Nicomachean Ethics*. The anonymous commentary on Books 2, 3, 4 and 5 of the *Nicomachean Ethics*, in *CAG* vol. 20, is derived from Adrastus, a generation later.[3] The commentaries by Alexander of Aphrodisias (appointed to his chair between A.D. 198 and 209) represent the fullest flowering of the Aristotelian tradition. To his successors Alexander was The Commentator *par excellence*. To give but one example (not from a commentary) of his skill at defending and elaborating Aristotle's views, one might refer to his defence of Aristotle's claim that space is finite against the objection that an edge of space is conceptually problematic.[4] Themistius (*fl.* late 340s to 384 or 385) saw himself as the inventor of paraphrase, wrongly thinking that the job of commentary was completed.[5] In fact, the Neoplatonists were to introduce new dimensions into commentary. Themistius' own relation to the Neoplatonist as opposed to the Aristotelian tradition is a matter of controversy,[6] but it would be

[3] Anthony Kenny, *The Aristotelian Ethics* (Oxford 1978), 37, n.3; Paul Moraux, *Der Aristotelismus bei den Griechen*, vol. 2 (Berlin 1984), 323-30.

[4] Alexander, *Quaestiones* 3.12, discussed in my *Matter, Space and Motion* (London and Ithaca, N.Y. 1988). For Alexander see R.W. Sharples, 'Alexander of Aphrodisias: scholasticism and innovation', in W. Haase (ed.), *Aufstieg und Niedergang der römischen Welt*, part 2 *Principat*, vol. 36.2, *Philosophie und Wissenschaften* (1987).

[5] Themistius *in An. Post.* 1,2-12. See H.J. Blumenthal, 'Photius on Themistius (Cod.74): did Themistius write commentaries on Aristotle?', *Hermes* 107 (1979), 168-82.

[6] For different views, see H.J. Blumenthal, 'Themistius, the last Peripatetic commentator on Aristotle?', in Glen W. Bowersock, Walter Burkert, Michael C.J. Putnam, *Arktouros*, Hellenic Studies Presented to Bernard M.W. Knox (Berlin and

agreed that his commentaries show far less bias than the full-blown Neoplatonist ones. They are also far more informative than the designation 'paraphrase' might suggest, and it has been estimated that Philoponus' *Physics* commentary draws silently on Themistius six hundred times.[7] The pseudo-Alexandrian commentary on *Metaphysics* 6–14, of unknown authorship, has been placed by some in the same group of commentaries as being earlier than the fifth century.[8]

By far the largest group of extant commentaries is that of the Neoplatonists up to the sixth century A.D. Nearly all the major Neoplatonists, apart from Plotinus (the founder of Neoplatonism), wrote commentaries on Aristotle, although those of Iamblichus (*c.* 250 – *c.* 325) survive only in fragments, and those of three Athenians, Plutarchus (died 432), his pupil Proclus (410 – 485) and the Athenian Damascius (*c.* 462 – after 538), are lost.[9] As a result of these losses, most of the extant Neoplatonist commentaries come from the late fifth and the sixth centuries and a good proportion from Alexandria. There are commentaries by Plotinus' disciple and editor Porphyry (232 – 309), by Iamblichus' pupil Dexippus (*c.* 330), by Proclus' teacher Syrianus (died *c.* 437), by Proclus' pupil Ammonius (435/445 – 517/526), by Ammonius' three pupils Philoponus (*c.* 490 to 570s), Simplicius (wrote after 532, probably after 538) and Asclepius (sixth century), by Ammonius' next but one successor Olympiodorus (495/505 – after 565), by Elias (*fl.* 541?), by David (second half of the sixth century, or beginning of the seventh) and by Stephanus (took the chair in Constantinople *c.* 610). Further,

N.Y., 1979), 391-400; E.P. Mahoney, 'Themistius and the agent intellect in James of Viterbo and other thirteenth-century philosophers: (Saint Thomas Aquinas, Siger of Brabant and Henry Bate)', *Augustiniana* 23 (1973), 422-67, at 428-31; id., 'Neoplatonism, the Greek commentators and Renaissance Aristotelianism', in D.J. O'Meara (ed.), *Neoplatonism and Christian Thought* (Albany N.Y. 1982), 169-77 and 264-82, esp. n. 1, 264-6; Robert Todd, introduction to translation of Themistius *in DA 3,4-8,* forthcoming in a collection of translations by Frederick Schroeder and Robert Todd of material in the commentators relating to the intellect.

[7] H. Vitelli, *CAG* 17, p. 992, s.v. Themistius.

[8] The similarities to Syrianus (died *c.*437) have suggested to some that it predates Syrianus (most recently Leonardo Tarán, review of Paul Moraux, *Der Aristotelismus*, vol. 1, in *Gnomon* 46 (1981), 721-50 at 750), to others that it draws on him (most recently P. Thillet, in the Budé edition of Alexander *de Fato*, p. lvii). Praechter ascribed it to Michael of Ephesus (eleventh or twelfth century), in his review of *CAG* 22.2, in *Göttingische Gelehrte Anzeiger* 168 (1906), 861-907.

[9] The Iamblichus fragments are collected in Greek by Bent Dalsgaard Larsen, *Jamblique de Chalcis, Exégète et Philosophe* (Aarhus 1972), vol.2. Most are taken from Simplicius, and will accordingly be translated in due course. The evidence on Damascius' commentaries is given in L.G. Westerink, *The Greek Commentaries on Plato's Phaedo*, vol.2., Damascius (Amsterdam 1977), 11-12; on Proclus' in L.G. Westerink, *Anonymous Prolegomena to Platonic Philosophy* (Amsterdam 1962), xii, n.22; on Plutarchus' in H.M. Blumenthal, 'Neoplatonic elements in the de Anima commentaries', *Phronesis* 21 (1976), 75.

a commentary on the *Nicomachean Ethics* has been ascribed to Heliodorus of Prusa, an unknown pre-fourteenth-century figure, and there is a commentary by Simplicius' colleague Priscian of Lydia on Aristotle's successor Theophrastus. Of these commentators some of the last were Christians (Philoponus, Elias, David and Stephanus), but they were Christians writing in the Neoplatonist tradition, as was also Boethius who produced a number of commentaries in Latin before his death in 525 or 526.

The third group comes from a much later period in Byzantium. The Berlin edition includes only three out of more than a dozen commentators described in Hunger's *Byzantinisches Handbuch*.[10] The two most important are Eustratius (1050/1060 – *c*. 1120), and Michael of Ephesus. It has been suggested that these two belong to a circle organised by the princess Anna Comnena in the twelfth century, and accordingly the completion of Michael's commentaries has been redated from 1040 to 1138.[11] His commentaries include areas where gaps had been left. Not all of these gap-fillers are extant, but we have commentaries on the neglected biological works, on the *Sophistici Elenchi*, and a small fragment of one on the *Politics*. The lost *Rhetoric* commentary had a few antecedents, but the *Rhetoric* too had been comparatively neglected. Another product of this period may have been the composite commentary on the *Nicomachean Ethics* (*CAG* 20) by various hands, including Eustratius and Michael, along with some earlier commentators, and an improvisation for Book 7. Whereas Michael follows Alexander and the conventional Aristotelian tradition, Eustratius' commentary introduces Platonist, Christian and anti-Islamic elements.[12]

The composite commentary was to be translated into Latin in the next century by Robert Grosseteste in England. But Latin translations of various logical commentaries were made from the Greek still earlier by James of Venice (*fl. c.* 1130), a contemporary of Michael of Ephesus, who may have known him in Constantinople.

[10] Herbert Hunger, *Die hochsprachliche profane Literatur der Byzantiner*, vol.1 (= *Byzantinisches Handbuch*, part 5, vol.1) (Munich 1978), 25-41. See also B.N. Tatakis, *La Philosophie Byzantine* (Paris 1949).

[11] R. Browning, 'An unpublished funeral oration on Anna Comnena', *Proceedings of the Cambridge Philological Society* n.s. 8 (1962), 1-12, esp. 6-7.

[12] R. Browning, op. cit. H.D.P. Mercken, *The Greek Commentaries of the Nicomachean Ethics of Aristotle in the Latin Translation of Grosseteste, Corpus Latinum Commentariorum in Aristotelem Graecorum* VI 1 (Leiden 1973), ch.1, 'The compilation of Greek commentaries on Aristotle's Nicomachean Ethics'. Sten Ebbesen, 'Anonymi Aurelianensis I Commentarium in *Sophisticos Elenchos', Cahiers de l'Institut Moyen Age Grecque et Latin* 34 (1979), 'Boethius, Jacobus Veneticus, Michael Ephesius and "Alexander" ', pp. v-xiii; id., *Commentators and Commentaries on Aristotle's Sophistici Elenchi*, 3 parts, *Corpus Latinum Commentariorum in Aristotelem Graecorum*, vol. 7 (Leiden 1981); A. Preus, *Aristotle and Michael of Ephesus on the Movement and Progression of Animals* (Hildesheim 1981), introduction.

And later in that century other commentaries and works by commentators were being translated from Arabic versions by Gerard of Cremona (died 1187).[13] So the twelfth century resumed the transmission which had been interrupted at Boethius' death in the sixth century.

The Neoplatonist commentaries of the main group were initiated by Porphyry. His master Plotinus had discussed Aristotle, but in a very independent way, devoting three whole treatises (*Enneads* 6.1–3) to attacking Aristotle's classification of the things in the universe into categories. These categories took no account of Plato's world of Ideas, were inferior to Plato's classifications in the *Sophist* and could anyhow be collapsed, some of them into others. Porphyry replied that Aristotle's categories could apply perfectly well to the world of intelligibles and he took them as in general defensible.[14] He wrote two commentaries on the *Categories*, one lost, and an introduction to it, the *Isagôgê*, as well as commentaries, now lost, on a number of other Aristotelian works. This proved decisive in making Aristotle a necessary subject for Neoplatonist lectures and commentary. Proclus, who was an exceptionally quick student, is said to have taken two years over his Aristotle studies, which were called the Lesser Mysteries, and which preceded the Greater Mysteries of Plato.[15] By the time of Ammonius, the commentaries reflect a teaching curriculum which begins with Porphyry's *Isagôgê* and Aristotle's *Categories*, and is explicitly said to have as its final goal a (mystical) ascent to the supreme Neoplatonist deity, the One.[16] The curriculum would have progressed from Aristotle to Plato, and would have culminated in Plato's *Timaeus* and *Parmenides*. The latter was read as being about the One, and both works were established in this place in the curriculum at least by

[13] For Grosseteste, see Mercken as in n. 12. For James of Venice, see Ebbesen as in n. 12, and L. Minio-Paluello, 'Jacobus Veneticus Grecus', *Traditio* 8 (1952), 265-304; id., 'Giacomo Veneto e l'Aristotelismo Latino', in Pertusi (ed.), *Venezia e l'Oriente fra tardo Medioevo e Rinascimento* (Florence 1966), 53-74, both reprinted in his *Opuscula* (1972). For Gerard of Cremona, see M. Steinschneider, *Die europäischen Übersetzungen aus dem arabischen bis Mitte des 17. Jahrhunderts* (repr. Graz 1956); E. Gilson, *History of Christian Philosophy in the Middle Ages* (London 1955), 235-6 and more generally 181-246. For the translators in general, see Bernard G. Dod, 'Aristoteles Latinus', in N. Kretzmann, A. Kenny, J. Pinborg (eds). *The Cambridge History of Latin Medieval Philosophy* (Cambridge 1982).

[14] See P. Hadot, 'L'harmonie des philosophies de Plotin et d'Aristote selon Porphyre dans le commentaire de Dexippe sur les Catégories', in *Plotino e il neoplatonismo in Oriente e in Occidente* (Rome 1974), 31-47; A.C. Lloyd, 'Neoplatonic logic and Aristotelian logic', *Phronesis* 1 (1955-6), 58-79 and 146-60.

[15] Marinus, *Life of Proclus* ch.13, 157,41 (Boissonade).

[16] The introductions to the *Isagôgê* by Ammonius, Elias and David, and to the *Categories* by Ammonius, Simplicius, Philoponus, Olympiodorus and Elias are discussed by L.G. Westerink, *Anonymous Prolegomena* and I. Hadot, 'Les Introductions', see n. 2. above.

the time of Iamblichus, if not earlier.[17]

Before Porphyry, it had been undecided how far a Platonist should accept Aristotle's scheme of categories. But now the proposition began to gain force that there was a harmony between Plato and Aristotle on most things.[18] Not for the only time in the history of philosophy, a perfectly crazy proposition proved philosophically fruitful. The views of Plato and of Aristotle had both to be transmuted into a new Neoplatonist philosophy in order to exhibit the supposed harmony. Iamblichus denied that Aristotle contradicted Plato on the theory of Ideas.[19] This was too much for Syrianus and his pupil Proclus. While accepting harmony in many areas,[20] they could see that there was disagreement on this issue and also on the issue of whether God was causally responsible for the existence of the ordered physical cosmos, which Aristotle denied. But even on these issues, Proclus' pupil Ammonius was to claim harmony, and, though the debate was not clear cut,[21] his claim was on the whole to prevail. Aristotle, he maintained, accepted Plato's Ideas,[22] at least in the form of principles (*logoi*) in the divine intellect, and these principles were in turn causally responsible for the beginningless existence of the physical universe. Ammonius wrote a whole book to show that Aristotle's God was thus an efficient cause, and though the book is lost, some of its principal arguments are preserved by Simplicius.[23] This tradition helped to make it possible for Aquinas to claim Aristotle's God as a Creator, albeit not in the sense of giving

[17] Proclus *in Alcibiadem 1* p.11 (Creuzer); Westerink, *Anonymous Prolegomena*, ch. 26, 12f. For the Neoplatonist curriculum see Westerink, Festugière, P. Hadot and I. Hadot in n. 2.

[18] See e.g. P. Hadot (1974), as in n. 14 above; H.J. Blumenthal, 'Neoplatonic elements in the de Anima commentaries', *Phronesis* 21 (1976), 64-87; H.A. Davidson, 'The principle that a finite body can contain only finite power', in S. Stein and R. Loewe (eds), *Studies in Jewish Religious and Intellectual History presented to A. Altmann* (Alabama 1979), 75-92; Carlos Steel, 'Proclus et Aristote', Proceedings of the Congrès Proclus held in Paris 1985, J. Pépin and H.D. Saffrey (eds), *Proclus, lecteur et interprète des anciens* (Paris 1987), 213-25; Koenraad Verrycken, *God en Wereld in de Wijsbegeerte van Ioannes Philoponus*, Ph.D. Diss. (Louvain 1985).

[19] Iamblichus ap. Elian *in Cat.* 123,1-3.

[20] Syrianus *in Metaph.* 80,4-7; Proclus *in Tim.* 1.6,21-7,16.

[21] Asclepius sometimes accepts Syranius' interpretation (*in Metaph.* 433,9-436,6); which is, however, qualified, since Syrianus thinks Aristotle is really committed willy-nilly to much of Plato's view (*in Metaph.* 117,25-118,11; ap. Asclepium *in Metaph.* 433,16; 450,22); Philoponus repents of his early claim that Plato is not the target of Aristotle's attack, and accepts that Plato is rightly attacked for treating ideas as independent entities outside the divine Intellect (*in DA* 37,18-31; *in Phys.* 225,4-226,11; *contra Procl.* 26,24-32,13; *in An. Post.* 242,14–243,25).

[22] Asclepius *in Metaph* from the voice of (i.e. from the lectures of) Ammonius 69,17-21; 71,28; cf. Zacharias *Ammonius, Patrologia Graeca* vol. 85, col. 952 (Colonna).

[23] Simplicius *in Phys.* 1361,11-1363,12. See H.A. Davidson; Carlos Steel; Koenraad Verrycken in n.18 above.

the universe a beginning, but in the sense of being causally responsible for its beginningless existence.[24] Thus what started as a desire to harmonise Aristotle with Plato finished by making Aristotle safe for Christianity. In Simplicius, who goes further than anyone,[25] it is a formally stated duty of the commentator to display the harmony of Plato and Aristotle in most things.[26] Philoponus, who with his independent mind had thought better of his earlier belief in harmony, is castigated by Simplicius for neglecting this duty.[27]

The idea of harmony was extended beyond Plato and Aristotle to Plato and the Presocratics. Plato's pupils Speusippus and Xenocrates saw Plato as being in the Pythagorean tradition.[28] From the third to first centuries B.C., pseudo-Pythagorean writings present Platonic and Aristotelian doctrines as if they were the ideas of Pythagoras and his pupils,[29] and these forgeries were later taken by the Neoplatonists as genuine. Plotinus saw the Presocratics as precursors of his own views,[30] but Iamblichus went far beyond him by writing ten volumes on Pythagorean philosophy.[31] Thereafter Proclus sought to unify the whole of Greek philosophy by presenting it as a continuous clarification of divine revelation,[32] and Simplicius argued for the same general unity in order to rebut Christian charges of contradictions in pagan philosophy.[33]

Later Neoplatonist commentaries tend to reflect their origin in a teaching curriculum:[34] from the time of Philoponus, the discussion is often divided up into lectures, which are subdivided into studies of doctrine and of text. A general account of Aristotle's philosophy is prefixed to the *Categories* commentaries and divided, according to a formula of Proclus,[35] into ten questions. It is here that commentators explain the eventual purpose of studying Aristotle (ascent to the One) and state (if they do) the requirement of

[24] See Richard Sorabji, *Matter, Space and Motion* (London and Ithaca N.Y. 1988), ch. 15.

[25] See e.g. H.J. Blumenthal in n. 18 above.

[26] Simplicius *in Cat.* 7,23-32.

[27] Simplicius *in Cael.* 84,11-14; 159,2-9. On Philoponus' *volte face* see n. 21 above.

[28] See e.g. Walter Burkert, *Weisheit und Wissenschaft* (Nürnberg 1962), translated as *Lore and Science in Ancient Pythagoreanism* (Cambridge Mass. 1972), 83-96.

[29] See Holger Thesleff, *An Introduction to the Pythagorean writings of the Hellenistic Period* (Åbo 1961); Thomas Alexander Szlezák, *Pseudo-Archytas über die Kategorien*, Peripatoi vol. 4 (Berlin and New York 1972).

[30] Plotinus e.g. 4.8.1; 5.1.8 (10-27); 5.1.9.

[31] See Dominic O'Meara, *Pythagoras Revived: Mathematics and Philosophy in late Antiquity* (Oxford 1989).

[32] See Christian Guérard, 'Parménide d'Elée selon les Néoplatoniciens', forthcoming.

[33] Simplicius *in Phys.* 28,32-29,5; 640,12-18. Such thinkers as Epicurus and the Sceptics, however, were not subject to harmonisation.

[34] See the literature in n. 2 above.

[35] ap. Elian *in Cat.* 107,24-6.

displaying the harmony of Plato and Aristotle. After the ten-point introduction to Aristotle, the *Categories* is given a six-point introduction, whose antecedents go back earlier than Neoplatonism, and which requires the commentator to find a unitary theme or scope (*skopos*) for the treatise. The arrangements for late commentaries on Plato are similar. Since the Plato commentaries form part of a single curriculum they should be studied alongside those on Aristotle. Here the situation is easier, not only because the extant corpus is very much smaller, but also because it has been comparatively well served by French and English translators.[36]

Given the theological motive of the curriculum and the pressure to harmonise Plato with Aristotle, it can be seen how these commentaries are a major source for Neoplatonist ideas. This in turn means that it is not safe to extract from them the fragments of the Presocratics, or of other authors, without making allowance for the Neoplatonist background against which the fragments were originally selected for discussion. For different reasons, analogous warnings apply to fragments preserved by the pre-Neoplatonist commentator Alexander.[37] It will be another advantage of the present translations that they will make it easier to check the distorting effect of a commentator's background.

Although the Neoplatonist commentators conflate the views of Aristotle with those of Neoplatonism, Philoponus alludes to a certain convention when he quotes Plutarchus expressing disapproval of Alexander for expounding his own philosophical doctrines in a commentary on Aristotle.[38] But this does not stop Philoponus from later inserting into his own commentaries on the *Physics* and *Meteorology* his arguments in favour of the Christian view of Creation. Of course, the commentators also wrote independent works of their own, in which their views are expressed independently of the exegesis of Aristotle. Some of these independent works will be included in the present series of translations.

The distorting Neoplatonist context does not prevent the commentaries from being incomparable guides to Aristotle. The

[36] English: Calcidius *in Tim.* (parts by van Winden; den Boeft); Iamblichus fragments (Dillon); Proclus *in Tim.* (Thomas Taylor); Proclus *in Parm.* (Dillon); Proclus *in Parm.*, end of 7th book, from the Latin (Klibansky, Labowsky, Anscombe); Proclus *in Alcib. 1* (O'Neill); Olympiodorus and Damascius *in Phaedonem* (Westerink); Damascius *in Philebum* (Westerink); *Anonymous Prolegomena to Platonic Philosophy* (Westerink). See also extracts in Thomas Taylor, *The Works of Plato*, 5 vols. (1804). French: Proclus *in Tim.* and *in Rempublicam* (Festugière); *in Parm.* (Chaignet); Anon. *in Parm.* (P. Hadot); Damascius *in Parm.* (Chaignet).

[37] For Alexander's treatment of the Stoics, see Robert B. Todd, *Alexander of Aphrodisias on Stoic Physics* (Leiden 1976), 24-9.

[38] Philoponus *in DA* 21,20-3.

introductions to Aristotle's philosophy insist that commentators must have a minutely detailed knowledge of the entire Aristotelian corpus, and this they certainly have. Commentators are also enjoined neither to accept nor reject what Aristotle says too readily, but to consider it in depth and without partiality. The commentaries draw one's attention to hundreds of phrases, sentences and ideas in Aristotle, which one could easily have passed over, however often one read him. The scholar who makes the right allowance for the distorting context will learn far more about Aristotle than he would be likely to on his own.

The relations of Neoplatonist commentators to the Christians were subtle. Porphyry wrote a treatise explicitly against the Christians in 15 books, but an order to burn it was issued in 448, and later Neoplatonists were more circumspect. Among the last commentators in the main group, we have noted several Christians. Of these the most important were Boethius and Philoponus. It was Boethius' programme to transmit Greek learning to Latin-speakers. By the time of his premature death by execution, he had provided Latin translations of Aristotle's logical works, together with commentaries in Latin but in the Neoplatonist style on Porphyry's *Isagôgê* and on Aristotle's *Categories* and *de Interpretatione*, and interpretations of the *Prior* and *Posterior Analytics, Topics* and *Sophistici Elenchi*. The interruption of his work meant that knowledge of Aristotle among Latin-speakers was confined for many centuries to the logical works. Philoponus is important both for his proofs of the Creation and for his progressive replacement of Aristotelian science with rival theories, which were taken up at first by the Arabs and came fully into their own in the West only in the sixteenth century.

Recent work has rejected the idea that in Alexandria the Neoplatonists compromised with Christian monotheism by collapsing the distinction between their two highest deities, the One and the Intellect. Simplicius (who left Alexandria for Athens) and the Alexandrians Ammonius and Asclepius appear to have acknowledged their beliefs quite openly, as later did the Alexandrian Olympiodorus, despite the presence of Christian students in their classes.[39]

The teaching of Simplicius in Athens and that of the whole pagan Neoplatonist school there was stopped by the Christian Emperor Justinian in 529. This was the very year in which the Christian

[39] For Simplicius, see I. Hadot, *Le Problème du Néoplatonisme Alexandrin: Hiéroclès et Simplicius* (Paris 1978); for Ammonius and Asclepius, Koenraad Verrycken, *God en Wereld in de Wijsbegeerte van Ioannes Philoponus*, Ph.D. Diss. (Louvain 1985); for Olympiodorus, L.G. Westerink, *Anonymous Prolegomena to Platonic Philosophy* (Amsterdam 1962).

Philoponus in Alexandria issued his proofs of Creation against the earlier Athenian Neoplatonist Proclus. Archaeological evidence has been offered that, after their temporary stay in Ctesiphon (in present-day Iraq), the Athenian Neoplatonists did not return to their house in Athens, and further evidence has been offered that Simplicius went to Ḥarrān (Carrhae), in present-day Turkey near the Iraq border.[40] Wherever he went, his commentaries are a treasure house of information about the preceding thousand years of Greek philosophy, information which he painstakingly recorded after the closure in Athens, and which would otherwise have been lost. He had every reason to feel bitter about Christianity, and in fact he sees it and Philoponus, its representative, as irreverent. They deny the divinity of the heavens and prefer the physical relics of dead martyrs.[41] His own commentaries by contrast culminate in devout prayers.

Two collections of articles by various hands have been published, to make the work of the commentators better known. The first is devoted to Philoponus;[42] the second is about the commentators in general, and goes into greater detail on some of the issues briefly mentioned here.[43]

[40] Alison Frantz, 'Pagan philosophers in Christian Athens', *Proceedings of the American Philosophical Society* 119 (1975), 29-38; M. Tardieu, 'Témoins orientaux du *Premier Alcibiade* à Ḥarrān et à Nag 'Hammādi', *Journal Asiatique* 274 (1986); id., 'Les calendriers en usage à Ḥarrān d'après les sources arabes et le commentaire de Simplicius à la *Physique* d'Aristote', in I. Hadot (ed.), *Simplicius, sa vie, son oeuvre, sa survie* (Berlin 1987), 40-57; *Coutumes nautiques mésopotamiennes chez Simplicius*, in preparation. The opposing view that Simplicius returned to Athens is most fully argued by Alan Cameron, 'The last days of the Academy at Athens', *Proceedings of the Cambridge Philological Society* 195, n.s. 15 (1969), 7-29.

[41] Simplicius *in Cael.* 26,4-7; 70,16-18; 90,1-18; 370,29-371,4. See on his whole attitude Philippe Hoffmann, 'Simplicius' polemics', in Richard Sorabji (ed.), *Philoponus and the Rejection of Aristotelian Science* (London and Ithaca, N.Y. 1987).

[42] Richard Sorabji (ed.), *Philoponus and the Rejection of Aristotelian Science* (London and Ithaca, N.Y. 1987).

[43] Richard Sorabji (ed.), *Aristotle Transformed: the ancient commentators and their influence* (London and Ithaca, N.Y. 1990). The lists of texts and previous translations of the commentaries included in Wildberg, *Philoponus Against Aristotle on the Eternity of the World* (pp.12ff.) are not included here. The list of translations should be augmented by: F.L.S. Bridgman, Heliodorus (?) in *Ethica Nicomachea*, London 1807.

I am grateful for comments to Henry Blumenthal, Victor Caston, I. Hadot, Paul Mercken, Alain Segonds, Robert Sharples, Robert Todd, L.G. Westerink and Christian Wildberg.

English-Greek Glossary

account: *logos*
actuality, actualisation: *energeia*
assertion, affirmation: *kataphasis*
attribute, accident: *sumbebêkos*, *pathos*

being, see essence

capacity, faculty: *dunamis*
category, predicate: *katêgoria*
cause: *aitia*, *aition*
change, motion: *kinêsis*
coherence: *akolouthia, allêloukhia*
concept, see thought
contradiction, opposition: *enantiôsis*
contrary, opposite: *enantion*

definition (see also account): *horos*, *logos*
differentia: *diaphora*
disposition: *diathesis*
division: *diairesis*

essence, existence, being, substance, subsistence, reality: *ousia*, *huparxis*, *hupostasis*, *to ti ên einai*

form: *eidos*

generation: *genesis*

homonym: *homônumos*

incorporeal: *asômatos*
intelligible: *noêtos*

knowledge, science: *epistêmê*, *gnôsis*

language, see speech

matter: *hulê*

motion, see change

name, word: *onoma*
negation: *apophasis*
number: *arithmos*

opposite, see contrary

part: *meros*
particular: *kath' hekaston*
potentiality: *dunamis*
predicate, see category
proprium, property: *idion*

quality: *poiotês, poion*
quantity: *posotês, poson*

relation: *pros ti*

science, see knowledge
sentence, see speech
sign, signification, reference: *sêmasia, sêmeion*
species: *eidos*
speech, language, utterance, sentence: *lexis, logos, phônê*
state: *hexis*
subject: *hupokeimenon*
substance: *ousia*

thing: *pragma*
thought, concept: *dianoia, ennoia, epinoia, noêma* (as process); *dianoêsis, noêsis*

universal: *katholon*

whole: *holon*
word, see name

Greek-English Index

References are to the page and line numbers of Busse's *CAG* edition, which appear in the margin of this translation.

Subject Index

References are to the page and line numbers of Busse's *CAG* edition, which appear in the margin of this translation.